COURSE OF MATHEMATICAL LOGIC

VOLUME I

SYNTHESE LIBRARY

MONOGRAPHS ON EPISTEMOLOGY,

LOGIC, METHODOLOGY, PHILOSOPHY OF SCIENCE,

SOCIOLOGY OF SCIENCE AND OF KNOWLEDGE,

AND ON THE MATHEMATICAL METHODS OF

SOCIAL AND BEHAVIORAL SCIENCES

ROLAND FRAÏSSÉ

COURSE OF MATHEMATICAL LOGIC

VOLUME I

Relation and Logical Formula

D. REIDEL PUBLISHING COMPANY
DORDRECHT-HOLLAND / BOSTON-U.S.A.

COURS DE LOGIQUE MATHÉMATIQUE, TOME I

First published by Gauthier-Villars and E. Nauwelaerts, Paris, Louvain, 1967

Second revised and improved edition, Gauthier-Villars, Paris, 1971

Translation edited by David Louvish

Library of Congress Catalog Card Number 72–95893

ISBN 90 277 0268 3

Published by D. Reidel Publishing Company,
P.O. Box 17, Dordrecht, Holland

Sold and distributed in the U.S.A., Canada, and Mexico
by D. Reidel Publishing Company, Inc.
306 Dartmouth Street, Boston,
Mass. 02116, U.S.A.

Printed in The Netherlands by D. Reidel, Dordrecht

TABLE OF CONTENTS

CHAPTER 3/RELATION, MULTIRELATION; OPERATOR AND PREDICATE

CHAPTER 4/LOCAL ISOMORPHISM; FREE OPERATOR AND FREE FORMULA

CHAPTER 5/FORMULA, OPERATOR, LOGICAL CLASS AND LOGICAL EQUIVALENCE; DENUMERABLE-MODEL THEOREM

CHAPTER 6. COMPLETENESS AND INTERPOLATION THEOREMS

CHAPTER 7. INTERPRETABILITY OF RELATIONS

PREFACE

This book is intended primarily for researchers specializing in mathematical logic; it may also serve the needs of advanced students desirous of embarking on research in logic. Volume 1 (at least) is also addressed to university and high-school students, mathematicians (not necessarily logicians) and philosophers, interested in a more rigorous notion of deductive reasoning.

The present two-volume edition reproduces and considerably expands the material of the one-volume first [French] edition published in 1967 (*Relation, Formule logique, Compacité, Complétude*). The new Volume 1, *Relation and Logical Formula*, contains the previous Chapters 1, 2, 3, 4, 5 and 8. Volume 2, *Model Theory*, reproduces the original Chapters 6 (Logical restriction, compactness) and 7 (Theory, axiomatic theories), together with five new chapters.

A summary of Volumes 1 and 2 is given below at the beginning of the Introduction. For the reader already familiar with the first edition, we mention the principal innovations in Volume 1. In Chapter 2 we have added a rather hasty outline of the proof of Post's theorem stating that every set of connections [connectives] closed under formulation is generated by a finite number of connections; this theorem is probably the most profound result to date in connective (alias propositional) logic. Chapter 4 now includes a study of the universal classes of Tarski and Vaught. Chapter 7, which is based mainly on the end of Chapter 5 of the first edition, also includes Julia Robinson's theorem on the interpretability [definability] of the set of integers in the field of rational numbers. Some of the exercises are entirely new; these are Exercises 3 and 9 in Chapter 3, Exercises 6, 7, 8 in Chapter 4, Exercises 4, 5, 6 in Chapter 5, Exercises 4, 5, 6, in Chapter 6, Exercises 2 and 3 in Chapter 7.

INTRODUCTION

Apart from certain refinements, the present book reproduces my courses taught since 1962 at the Faculté des Sciences in Paris, and subsequently at the Universities of Provence and Paris-6.

Our standpoint here is that of Semantics, the science of the connections between logical formulas and the relations or models that satisfy them, in contradistinction to Syntax, the science of the structure of logical formulas and of formal deduction.

In Volume 1, we assume the reader acquainted with the elements of combinatorial theory for our study of formulas and connections (Chapters 1 and 2), and with the elements of set theory for the sequel: algebra of relations (Chapter 3), logical calculus (Chapters 4 to 7); the most notable results of the latter are the denumerable-model theorem (Löwenheim-Skolem, Chapter 5), the completeness theorem (Gödel-Herbrand, Chapter 6), the interpolation theorem (Craig-Lyndon, Chapter 6), and the interpretability theorem (Beth, Chapter 7).

Volume 2 is devoted to Model Theory: compactness theorem, elimination of quantifiers, logical extension theorems of Mal'cev-Keisler and Henkin, Svenonius' interpretability theorem, axiomatizability of theories, ultraproducts, forcing, logic of infinite formulas.

To determine the position of our subject in the history of formal logic, now known as mathematical logic, we distinguish the following stages, from its birth to the present day (see the detailed bibliography of Church, 1936, updated in subsequent issues of the *Journal of Symbolic Logic*):

(1) Prehistory, extending from Aristotle up to the beginning of the 19th century, including the *ars combinatoria* of Leibniz, and the study of definitions by Pascal.

(2) From the forties of the 19th century to about 1915, the development of syntax (Boole, Frege, Peirce, Schröder) and axiomatic theory (Hilbert, Peano, Russell, Zermelo). Several fundamental mathematical theories were formalized: arithmetic, elementary geometry, analysis, set theory, and this was done in such a way as to preserve their consistency (non-

contradictoriness) and, in the successful cases, fruitfulness. Since Hilbert (1904), one is no longer content merely to confirm consistency experimentally; one tries to prove it, using a model, within the framework of some "reliable" classical theory. This is the beginning of semantics.

(3) From 1915 to about 1940, the period of mathematization. Löwenheim's denumerable-model theorem (1915), extended by Skolem (1920), would seem to be the earliest theorem of logic which is of mathematical interest. This theorem belongs to semantics; however, at this time syntax was also undergoing a renovation, and it gained in mathematical interest with the work of Post and Łukasiewicz (1921). The fundamental concept of a "model satisfied by a formula" used in all this work is not defined rigorously and generally until Tarski (1933; see references, 1936 and 1956). The completeness theorem was proved in 1930, via two very different approaches, by Gödel and Herbrand; subsequently, this gave rise to the compactness and interpolation theorems.

One might say that semantics stands in the same relation to syntax as does the theory of fields to the theory of algebraic equations. Instead of simply trying to determine whether a given logical formula is consistent or contradictory (a problem which nevertheless retains its importance and still presents enormous difficulties), one now lays siege to classes of relations satisfying a formula, classes of formulas satisfied by a relation, etc. As field theory establishes the impossibility of solving the equation of the fifth degree by radicals, so semantics establishes theorems of limitation or impossibility. Thus, the above-mentioned theorem, long known as the Löwenheim-Skolem paradox, precludes the possibility of a theory characterizing nondenumerable infinity. The compactness theorem states that we cannot construct a logical formula which is true for every finite set and false for the infinite sets. Above all, Gödel's insaturation [incompleteness] theorem (1931) obliges us to abandon all hope of an arithmetic based on a finite number of axioms or axiom schemata, in which every statement is either provable or refutable; Tarski (1936) shows that we cannot represent the truth or falsity of every proposition of a theory T within the theory T itself; Church (1936) destroys all hope for an algorithm or mechanical procedure which decides, for every proposition, whether it is a thesis or not.

(4) Since the forties of this century, a multitude of relationships between logic and algebra or arithmetic have come to light; these years

have seen the birth of algebraic logic, logical algebra, the theory of ultraproducts, recursive arithmetic, and so on. Above all, in axiomatic set theory we can now prove the consistency and independence of the axiom of choice and the continuum hypothesis (Gödel, 1938; see references, 1940; Cohen, 1963).

The above survey of concepts, with which the reader is assumed familiar, should indicate that we do not regard mathematical logic as a foundation, in the sense of a point of departure or introduction to mathematics. Our conception of the relationships between these two sciences will now be clarified.

If mathematics is supposed to include any science which develops by deductive methods from a small number of axioms, it must surely include logic, even if one excludes, say, mechanics and physical theories for the reason that they appeal to an overly large number of axioms, which are sometimes taken for granted. However, if one requires that a mathematical science serve as a tool for proving theorems of analysis or algebra, logic is only a candidate for inclusion in mathematics (note that the arithmetical Ax-Kochen theorem was obtained in 1965 by logical methods). Logic involves certain extra-mathematical problems, concerning adequacy or effectiveness, but this is also true of the "purest" theories. For example, does classical logic faithfully represent the usual reasoning of mathematicians? Can one derive all actually known theorems of arithmetic from Peano's arithmetic (1894), which specifies the axioms for addition and multiplication? Is the concept of recursive function a "correct" rigorous version of the experimental concept of a function computable by a machine?

Now that logic includes semantic theorems (that is to say, since about 1915), it would surely be unable to develop prior to mathematics, after the fashion of an introduction. For example, the denumerable-model theorem presupposes the definition of the set of integers and of relations constructed over the integers. Even if we reduce logic to syntax alone, we must define a "word" as a finite sequence of symbols, and a "formula" as a word constructed according to certain rules. Now the definition of a "finite sequence" of symbols presupposes the concept of a natural number; since the very beginnings of syntax, this involves inductive reasoning, sometimes disguised as follows: if the given property is true for atomic formulas; if its truth for a implies that it is true for "not a", and its truth for a and b implies that it is true for "a and b" and "a or b"; then the property is true for all connective [propositional] formulas in "not, and, or". We are thus obliged to base syntax on such strong axioms as those of arithmetic and combinatorial theory.

In our view, the position of logic is dictated by these considerations; the logician must begin with a fairly rich mathematical theory in order to achieve the representation of such concepts as word, formula, deduction, theory, axiomatics, consistence, saturation, etc. Then the exposition can refer to logic, which is simply the mathematical study of the above concepts. Naturally, the logician does not wait for a description of the rules of logic in order to reason correctly; these rules, to be set forth gradually, must already be applied when he phrases his first definition and proves his first theorem.

Though it is customary to compare the mathematician to a person playing a game, there is nevertheless an essential difference. The player must announce the rules of the game before beginning to play, for the sake of honesty, so that the opponent and witnesses will be able to check the legitimacy of his actions. Now the mathematician cannot provide his reader in advance with a means of checking the validity of his reasoning. A sufficiently intelligent person, though unacquainted with a game, should nevertheless be capable of understanding the rules by simply reading them. In mathematics, by contrast, exposition of the rules is part of the "game"; it is not an introduction to the game. One does not learn how to reason by reading a description of a mathematical formalism; one must in fact already know how to reason in order to understand a description of this type. Legitimacy of each action must be checked in mathematics quite as rigorously as in a game, but *after* each "play" has been made. The situation is similar to that of a dynasty or political regime, whose legitimacy can be established only after its founder has assumed power. Similarly, each of us can approve or disapprove of his actual existence only after it has been thrust upon him.

Only reluctantly do we abandon the ideal or illusion of logic as a foundation for mathematics. We are running the risk of a vicious circle: How can logic be a mere chapter in arithmetic if the latter can be axiomatized only within the framework of logic? Everything would be clearer if the laws of reasoning could be derived, or at least set forth, without any primitive concepts, starting from nothing. Our desire is unattainable, and our impasse in this respect seems to be a sequel to that of Descartes: the desire to attain truth by simple and indisputable reflection, with no point of departure other than ignorance or doubt.

Since logic fails, can we base mathematics on set theory, in the sense that every theory is the study of certain sets endowed with appropriate definitions? Even this is debatable. There exist set theories of vastly different natures. The axiom of choice may be accepted, or rejected either completely or with varying degrees of force. In Zermelo-Fraenkel set theory, every set may be regarded as an element; in von Neumann-Bernays-Gödel set theory, one needs sets, known as proper classes, which cannot be elements of any set. In the classical theory, there is only one empty set; in some theories (Fraenkel, 1925, or Mostowski, 1939, among others), one needs infinitely many empty sets, known as "*urelements*". If one admits the axiom of regularity (*Fundierungsaxiom*), every descending sequence in which each term is an element of its predecessor must terminate after finitely many terms in the empty set; but one must sometimes reject this axiom.

Above all, the classical theory of sets is merely a link in a chain of theories of ever-increasing richness. Even combinatorial theory may be regarded as a theory of finite sets. Analysis is included in a theory of finite and denumerable sets and sets whose cardinalities are either equal to or "slightly" greater than that of the continuum. The classical or Cantor theory encompasses only accessible sets, i.e., those built up from denumerable sets by forming power sets and unions. The needs of logicians, and recently those of algebraists, seem to dictate a theory of inaccessible sets; for the moment, these can be obtained by ensuring (in the axioms) that all the sets of the classical theory constitute a new set, to which one now applies the usual constructive procedures.

If everything in actual mathematics is a set, then the entities referred to by the same word "set", by an abuse of language which, though convenient, may nevertheless lead to confusion, exhibit ever-increasing differences. Thus, for a certain time, everything was a number, a convention which lent mathematics a certain impression of unity.

If one now wishes to relativize the meaning of the word "foundation", regarding it as a synonym for "representation of one theory in another", it should be clear that logic is largely the science of foundation in mathematics. The oldest example is that of the various geometries (including non-euclidean geometries), whose models have hitherto provided professional logicians with no additional inspiration; for example, Hilbert's axioms have given rise to more hopes than concrete results. The only major result in this direction is the decidability of elementary geometry, when the latter is assumed reduced to sentences expressible in terms of addition and multiplication of real numbers: each of these sentences is mechanically provable or refutable (Tarski, 1940, revised in 1951). Another example, which has received more attention, is the representation of a theory of sets including the axiom of choice in another set theory without this axiom (Gödel, 1940) and conversely (Cohen, 1963). As a third example, we may perhaps hope for a proof of the independence of arithmetics with or without the "Fermat conjecture", based on representation of one theory in another.

In general, one is obliged to base one theory on another whenever one wishes to prove consistency, non-deducibility of a statement from given axioms, independence of an axiom of other axioms, or undecidability of a theory (see especially Tarski, 1953). This field of activity, though by no means exhausting the range of modern mathematical logic, is a central preoccupation of many researchers, and is being followed with interest by all contemporary logicians.

I am indebted to M. René de Possel, Professor at the University of Paris-6, Director of the Laboratory of Advanced Research on Information Techniques at the CNRS, who has welcomed four preliminary versions of this book in the series *Logique à l'usage du calculateur*. Thanks are also due to workers at the Logic Seminar of the University of Provence at Marseilles, and at the attached seminars of Paris and Lyon (individual names will be mentioned in the text, in connection with the problems which they have solved or refined).

CHAPTER 1

WORD, FORMULA

In Chapters 1 and 2 we shall employ the axioms and definitions of combinatorial analysis, or the theory of finite sets. These may be, say, the Zermelo-Fraenkel axioms, which enable one to build up certain sets from the empty set: the axiom of extensionality (two sets having the same elements are identical); the existence of pair, union, and set of subsets (power set); the axiom of substitution (if for each element i of a a set b_i is given, then there exists a set whose elements are $b_i (i \in a)$); finally, the axiom of regularity (every nonempty set a contains an element disjoint from a). We shall not, however, need the axiom of infinity, and can even stipulate, without danger of contradiction, that every set will be finite; i.e., given a, every set of subsets of a will contain a minimal element under inclusion (Tarski's definition; see Exercise 1). We know that an integer is a finite ordinal; "a is an ordinal" means that $x \in y$ and $y \in a$ always imply $x \in a$, and that $x \in a$ and $y \in a$ always imply either $x \in y$, $x = y$, or $y \in x$ (see Exercises 2 and 3). Starting from $0 =$ the empty set, we get the integers $1 = \{0\}$, $2 = \{0, 1\}$, $3 = \{0, 1, 2\}, \ldots$; $a \leqslant b$ means $a \in b$ or $a = b$, and the integer $a + 1$ is defined as $a \cup \{a\}$. The ordered pair (a, b) is defined as the set whose elements are $\{a\}$ and $\{a, b\}$; the finite sequence of n terms $a, b, \ldots, h$ is the set whose elements are the [ordered] pairs $(1, a), (2, b), \ldots, (n, h)$.

1.1. WORD, RANK, OCCURRENCE, CONCATENATION, INTERVAL, SUBSTITUTION

Let E be a nonempty set. A *word* over E is a finite sequence U (with possible repetitions) of elements of E. U may be empty. A *term* of U is a pair (a, i) made up of an *element* a of E and its *rank* i in U; we shall write a_i instead of (a, i). The term a_i will be called an occurrence of a in U, or briefly an a in U.

Example. Let $\mathrm{E} = \{a, b, c, d\}$ and $\mathrm{U} = abdad$; then the terms a_1 and a_4 are two occurrences of a in U.

Given two words U and V, we define the *concatenation* of U and V, written UV, as the word obtained by setting down first the terms of U and then the terms of V (whose ranks are thus increased by the number of terms in U).

Concatenation, transforming U and V into UV, is associative; i.e., given words U, V, and W, we have (UV)W = U(VW). We shall write UVW; in general, given words U_i (where $i=1, 2, \ldots, h$), we obtain the word $U = U_1 U_2 \ldots U_h$. Each word U_i will be called an *interval* of U, where U_1 is an *initial* interval and U_h a *final* interval. The sequence of elements of $V = U_i$, indexed by their ranks in U, will be called an *occurrence* of V in U.

The word consisting of a single element a will be written as a instead of $\{a_1\}$, provided this does not lead to confusion. Thus, given a and b in E and words U and V over E, we get words $a\mathrm{U}b\mathrm{V}a$, $\mathrm{UV}aab\mathrm{V}$, etc.

1.1.1. Let us write a word in the form $V = UaU'$, where a is an element, i its rank, and U and U′ are two words; the *substitution of a word* A *for the occurrence* a_i *of* a is the operation transforming $V = UaU'$ into $W = UAU'$. The operation is denoted by $|^{A}_{a_i}$; the word W is called a *substitution instance* or *substitution result* of V, denoted by $W = V]^{A}_{a_i}$.

Let us write a word in the form $V = U_0 a U_1 a \ldots a U_h$, where a has no further occurrences in $U_0, U_1, \ldots, U_h$. The *substitution of a word* A *for the element* a is the operation transforming the given word V into the substitution instance $W = U_0 A U_1 A \ldots A U_h$; this will be denoted by $|^{A}_{a}$, and we write $W = V]^{A}_{a}$.

Given r elements $a_1, \ldots, a_r$, let us write a word in the form $V = U_0 a_i U_1 a_j \ldots a_k U_h$, where $i, j, \ldots, k$ are integers from $1, \ldots, r$, and the elements $a_1, \ldots, a_r$ have no further occurrences in $U_0, U_1, \ldots, U_h$. The *simultaneous substitution of r words* $A_1, \ldots, A_r$ *for* $a_1, \ldots, a_r$ is the operation transforming the given word V into the substitution instance $W = U_0 A_i U_1 A_j \ldots A_k U_h$. We shall denote this by $|^{A_1 \ldots A_r}_{a_1 \ldots a_r}$ and write $W = V]^{A_1 \ldots A_r}_{a_1 \ldots a_r}$.

1.2. Weight; Formula, Subformula, Occurrence of a Formula, Term Dominating Another Term

To each element a of E we associate an integer $h \geqslant 0$, called the *weight* of a [BOU, 1954, p. 15]. We define a *formula over* E as any nonempty word obtained by using the following two rules of construction, and by these rules only:

(1) *A word consisting of a single element of* E *of weight* 0 *is a formula.*

(2) *Given an element a of weight* $h \geqslant 1$ *and h formulas* $A_1, \ldots, A_h$, *the word* $aA_1 \ldots A_h$ *is a formula.*

Examples. If $a, b, c, d, \ldots$ are elements of weight 0, $a', b', \ldots$ are elements of weight 1, and $a'', b'', \ldots$ are elements of weight 2, then $a'b$, $a''cc$, $b''a'ba''cc$ are formulas.

We see that a formula always ends in an element of weight 0. If it does not reduce to a single element, then it begins with an element of positive weight.

When dealing only with elements of weights 0, 1 or 2 we shall often dispense with the theoretical notation just introduced and instead use the *parenthesis notation*, in which a'A (where a' is of weight 1) is written a'(A) and a''AB (where a'' is of weight 2) is written (A) a''(B); the formulas considered above will thus be written $a'b$, $ca''c$, and $(a'b)\, b''(ca''c)$. We shall nevertheless continue to say that the first term in the last formula is b'', or, more precisely, an occurrence of b'', the second term an a', the third a b, the fourth an a'', and the fifth and sixth are c's.

Note that if we substitute a formula for a term of weight 0 in some formula, we get a formula. See Exercise 5 for an examination of the substitution of elements or words for terms.

1.2.1. *If* A *is a formula and* U *a nonempty word, then* AU *is not a formula.*

▷ First, we may assume that A does not reduce to a single term a, since this would necessarily be of weight 0 and the word $AU = aU$ could not be a formula. Let us assume the assertion to be true for formulas with less than n terms, and show it to be true for a formula $A = aA_1 \ldots A_h$ with n terms, in which a is of weight h and $A_1, \ldots, A_h$ are formulas. Assume that $AU = aA_1 \ldots A_hU$ is a formula; since a is of weight h, there exist h formulas $B_1, \ldots, B_h$ such that $A_1 \ldots A_hU = B_1 \ldots B_h$. Then either B_1 is of the form A_1S, where S is a nonempty word, which is impossible, as A_1 has less than n terms; or A_1 is of the form B_1T, which is impossible, as $A_1 = B_1T$ means that B_1 has less than n terms. Therefore, $B_1 = A_1$. The same reasoning holds for A_2 and $B_2, \ldots$, up to A_h and B_h. ◁

Let $A_1, \ldots, A_h$ *and* $B_1, \ldots, B_k$ *be two sequences, of h and k formulas, respectively. If* $A_1 \ldots A_h = B_1 \ldots B_k$, *then* $h = k$ *and* $A_i = B_i$ *for each* $i = 1, 2, \ldots, h$.

Indeed, by 1.2.1 it is impossible to have $B_1 = A_1S$ or $A_1 = B_1T$ if S and T are nonempty words, so we must have $B_1 = A_1$. Similarly, we obtain $B_2 = A_2, \ldots, B_h = A_h$.

1.2.2. Given a formula A, we define a *subformula* B of A as any interval of A which is itself a formula. Alternatively: there exist an initial interval U and a final interval V such that A = UBV. Remember that, by 1.1, the sequence of elements in B, indexed by their ranks in A, is called an *occurrence* of B in A.

Examples. The formula A is itself a subformula of A. If $A = aA_1 \ldots A_h$, where a is of weight h, then the formulas $A_1, \ldots, A_h$ are subformulas of A which we shall call the *immediate* subformulas of A. The immediate subformulas of $A_1, \ldots, A_h$, if they exist, are subformulas of A. In 1.2.5 we shall see that every subformula of A is obtained by taking successive immediate subformulas.

Note that the same subformula B can have several occurrences in A; *example*: $A = (ca''c)\, b''(ca''c)$ and $B = ca''c$.

The relation "B is a subformula of A" is an ordering (reflexive, transitive, and antisymmetric).

We shall make use of the following obvious notions:

a formula occurrence *included* in another; for example, in $a'b''a'ba''cc$, or $a'((a'b)\, b''(ca''c))$, the occurrence of $a'b$ is included in that of $(a'b)\, b''(ca''c)$;

disjoint occurrences (without common term); same example – the occurrence of $a'b$ and the occurrence of $ca''c$ are disjoint;

consecutive occurrences; same example – the occurrence of $a'b$ and the occurrence of $ca''c$ are consecutive;

another example: in $b''b''a'ca'a'ca$, or $((a'c)\, b''(a'a'c))\, b''a$, the first occurrence of $a'c$, the occurrence of $a'a'c$ and the occurrence of a are consecutive, while the two occurrences of $a'c$ are disjoint but not consecutive.

1.2.3. *Let A be a formula and a a term of weight h; then the terms following a in the formula are distributed in a unique way among $\geqslant h$ occurrences of consecutive, disjoint formulas.*

▷ First, we may assume that A does not reduce to a single term a, as this would necessarily be of weight 0 and thus the assertion would be trivial. Now assume the statement to be true for formulas with less than n terms; we shall show that it holds for a formula $A = uA_1 \ldots A_k$ with n terms, of which the first u is of weight k. Now either a is the first term, whence $a = u$ and the assertion is obviously true, or a belongs to one of the A_i $(1 \leqslant i \leqslant k)$. As the latter has less than n terms, the assertion is assumed to be true for it, so the terms following a must be distributed among $\geqslant h$ occurrences of formulas, to which must be added the occurrences of $A_{i+1}, A_{i+2}, \ldots, A_k$.

It remains to show that the distribution of the terms following a among consecutive occurrences of formulas is unique; this is a consequence of 1.2.1. ◁

Example. In $b''b''a'ca'a'ca$ the second b'', of weight 2, is followed by the three consecutive formulas $a'c$, $a'a'c$, and a.

Every term in A *is the first term of a formula occurrence in* A (and, by 1.2.1, of only one formula occurrence).

It is sufficient to take the first h formula occurrences following the term.

We thus have a *bijective correspondence between formula occurrences in* A *and terms of* A, defined by associating each formula occurrence and its first term.

1.2.4. *Any two formula occurrences are either disjoint or one is included in the other.*

▷ Let S and T be two occurrences in A, not disjoint and such that the first term u of T, say, belongs to S. Now either u is of weight 0, so that T reduces to the single term u and is therefore included in S, or u is of weight $h \geqslant 1$. By 1.2.3, the terms following u in S are distributed in a unique way among at least h consecutive, disjoint occurrences. The term u followed by the first h occurrences constitutes a formula which, by 1.2.1, must be T, whence T is included in S. ◁

1.2.5. *If* B *is a subformula of* $A = aA_1 \ldots A_h$, *then either* B = A *or* B *is a subformula of one of* $A_1, \ldots, A_h$.

▷ Either the first term of the occurrence of B is a, whence B = A (see 1.2.1); or this first term is the first term of an A_i $(i = 1, \ldots, h)$, whence $B = A_i$; or, finally, it is a term of A_i other than the first, whence by 1.2.3, B is a subformula of A_i. ◁

It follows that a formula B is a *subformula* of A if B=A or if there exists a finite sequence of formulas, beginning with A and ending with B, in which each formula is an *immediate subformula* (see 1.2.2) of the preceding one.

Let us recall that if we have nonempty sets that are always either disjoint or included in each other, then the inclusion relation, denoted by $\leqslant$ instead of $\supseteq$, is a *ramified ordering* or *tree*, i.e. an ordering in which all elements preceding a given element are comparable (and thus constitute a chain or totally ordered set). From the above it thus follows that the *inclusion relation on the occurrences in a given formula is a tree.*

1.2.6. Given two terms t and u of A, we shall say that t *dominates* u if t is the first term in some subformula occurrence in A and u is any term in that occurrence. Thus, in the example $b'((a'c)\,b''(a'(a'c)))$, the occurrence of b' dominates that of b'', which in turn dominates the first and second occurrences of a', and so on.

Alternatively, we can say that t dominates u if the formula occurrence which begins with u is included in the formula occurrence beginning with t.

The relation "t dominates u" defined on the terms of A is a tree. Indeed, it is isomorphic to the inclusion relation on occurrences of subformulas of A: the mapping taking each term t onto the formula occurrence beginning with t is a bijection.

1.3. Weight of a word

We define the *weight* of a word U as the sum of weights of the terms of U, minus the number of terms, plus 1 [ROS, 1950]. Clearly, if U reduces to a single element a, its weight is that of a, and the weight of the empty word is obviously 1. If $p(\mathrm{U})$ and $p(\mathrm{V})$ are the weights of words U and V, then the weight of UV is $p(\mathrm{U})+p(\mathrm{V})-1$. In general, the weight of $\mathrm{U}_1 \ldots \mathrm{U}_h$ is $p(\mathrm{U}_1)+\cdots+p(\mathrm{U}_h)-h+1$. Note that the weight remains unchanged when we permute the terms in the word.

The *weight of a formula* is 0; this is obvious if the formula reduces to a single term, which must be of weight 0; otherwise, the formula A is $a\mathrm{A}_1 \ldots \mathrm{A}_h$, where a is of weight h, and the formulas $\mathrm{A}_1, \ldots, \mathrm{A}_h$ of weight 0, giving the value $h-(h+1)+1=0$.

1.3.1. *If* A *is a formula, the weight* $p(\mathrm{U})$ *of any initial interval* U *distinct from* A *is positive.*

▷ The assertion is true if A reduces to a single term. Let us assume it to be true for formulas with less than n terms and prove it for n. Let $\mathrm{A}=a\mathrm{A}_1\ldots\mathrm{A}_h$ have n terms, where a is of weight h and the A_i $(i=1,\ldots,h)$ are formulas having less than n terms. Thus $\mathrm{U}=a\mathrm{A}_1\ldots\mathrm{A}_k\mathrm{U}'$, where $k<h$ and U′ is an initial interval distinct from A_{k+1}; the weight $p(\mathrm{U}')$ is thus positive, by the induction hypothesis. Therefore,

$$p(\mathrm{U})=h+p(\mathrm{U}')-(k+2)+1=h-k-1+p(\mathrm{U}')\geqslant p(\mathrm{U}'),$$

so that $p(\mathrm{U})$ is positive. ◁

1.3.2. *If the word* U *is of negative weight, there exists an initial interval in* U *having weight* 0.

▷ Let us consider consecutive decompositions of U into a sequence of two words V and W. If V is empty, its weight is 1. As we go from a decomposition VW to V′W′, in which $\mathrm{V}'=\mathrm{V}$ followed by an element a and $\mathrm{W}'=\mathrm{W}$ minus its first term (occurrence of a), the weight $p(\mathrm{V})$ is replaced by $p(\mathrm{V}')=p(\mathrm{V})+p(a)-1$, and either increases, maintains its value, or, in case $p(a)=0$, decreases by 1. If we begin at weight 1, we can reach the negative weight $p(\mathrm{U})$ only by way of some V of weight 0. ◁

1.3.3. *A word* A *is a formula if and only if its weight is* 0 *and every initial interval in* A *is of positive weight.*

▷ If A is a formula, then, by 1.3 and 1.3.1, there is nothing further to prove. Conversely, assume that the conditions hold and that the assertion is true for words of less than n terms. Let A have n terms, the first of which, a, is of weight $h\geqslant 1$; then the word $\mathrm{A}'=\mathrm{A}$ minus its first term is of weight $p(\mathrm{A}')\leqslant 0$, since

$$h+p(\mathrm{A}')-1=0.$$

By 1.3.2 there exists a first word A_1 of weight 0 which is an initial interval of A′, so let us write $\mathrm{A}'=\mathrm{A}_1\mathrm{A}_1'$. If $h\geqslant 2$ then $p(\mathrm{A}_1')\leqslant 0$, since $h+p(\mathrm{A}_1')-2=0$ (because $\mathrm{A}=a\mathrm{A}_1\mathrm{A}_1'$). There thus exists a first word A_2 of weight 0 which is an initial interval of A_1'. Continuing in this way we get words $A_1, A_2, \ldots, A_h$, which have less than n terms and satisfy the conditions of the assertion; these are therefore formulas, and A has the form $aA_1 \ldots A_h$, which is a formula. ◁

1.4. ATOM; WEIGHTED SUBSTITUTION

Let us regard the positive integers i as our elements of weight 0, calling them *atoms* and denoting them by a_i (rather than i).

1.4.1. Let A be a formula, t *an occurrence* in A of an element of weight $h \geqslant 0$, and B a formula. We define the *weighted substitution instance* $A]_t^B$ as follows. Take the subformula of A which begins with t, say $tA_1 \ldots A_h$, where A_i $(i=1, \ldots, h)$ are subformulas of A; then there exist two words U and V such that $A = UtA_1 \ldots A_h V$. Substitute (in the ordinary sense of 1.1.1) the formula A_i for each occurrence in B of every atom a_i for which $i \leqslant h$, leaving the atoms for which $i > h$, if they exist, unchanged. Denote the formula thus obtained by B′, and set $A]_t^B = UB'V$.

Examples. Let $A = (aa''a'a)\, b''(a'b)$, and perform weighted substitution of $B = a_1c''a'a_1$ for the second occurrence of a'; we get $A_1 = b$, so $B' = bc''a'b$ and the substitution instance will be $(aa''a'a)\, b''(bc''a'b)$. Taking $B = ac''(a'a)$, where a is not an atom, we again have $A_1 = b$, but now $B' = B = ac''(a'a)$ and substitution gives $(aa''(a'a))\, b''(ac''(a'a))$.

Now let $A = (aa''c)\, b''b$ and substitute $B = a_2b''(a'a_2)$ for the a'', which is of weight 2; we get $A_1 = a$ and $A_2 = c$, whence $B' = cb''(a'c)$ and the substitution instance is thus $(cb''(a'c))\, b''b$. Starting with $A = (aa''c)\, b''b$ and substituting the same B, but now for b'', we get $A_1 = aa''c$ and $A_2 = b$, so $B' = bb''(a'b)$ and the substitution result coincides with B′. Substituting for b'' the formula $B = ab''(a'a)$, where a is not an atom, we get $B' = B = ab''(a'a)$. Finally, substituting $B = a_1b''(a'a_2)$ for b'', we get $A_1 = aa''c$ and $A_2 = b$, so that $B' = (aa''c)\, b''(a'b)$ is the desired substitution instance.

If the term t is of weight 0, this reduces to ordinary substitution of B for t.

If the term t is of weight h and B is a formula $\alpha a_1 \ldots a_h$, we get ordinary substitution of α for t.

If B does not contain any of the atoms $a_1, \ldots, a_h$, then $B' = B$; we can then say that we get the substitution of a formula B for the formula occurrence $tA_1 \ldots A_h$.

Finally, note that if t is the first term in A, then the substitution instance coincides with the B′ obtained from B by ordinary simultaneous substitution of $A_1, \ldots, A_h$ for $a_1, \ldots, a_h$.

1.4.2. Let A be a formula, b *an element* of weight $h \geqslant 0$ having one or more occurrences in A, and B a formula. We define the *weighted substitution* instance $A]_b^B$ as follows.

We write A in the form $tA_1 \ldots A_k$, where the first term t is of weight k and A_i $(i=1, \ldots, k)$ are the immediate subformulas. Now, either $k=0$ or $k \geqslant 1$. In the first case, i.e., $k=0$ and A reduces to its first term t, then if t is not an occurrence of b we set $A]_b^B = t$, and if t is a b we set $A]_b^B = B$. In the second case, i.e., $k \geqslant 1$, assume that the words $A_i]_b^B$ have been obtained; then if t is not a b we set $A]_b^B = t(A_1]_b^B) \ldots (A_k]_b^B)$, while if t is a b, so that $k=h$, then $A]_b^B$ is obtained from B by simultaneous substitution of the expression $A_i]_b^B$ (in the ordinary sense of 1.1.1) for each atom a_i $(i=1, \ldots, h)$ in B.

If b has weight zero, this operation reduces to ordinary substitution of B for b.

If b has only one occurrence t in A, this operation reduces to weighted substitution for the occurrence t (see 1.4.1).

We can also reduce matters to substitution for occurrences in the general case. We order the occurrences of b in a sequence $t_1, \ldots, t_r$, in which each occurrence which dominates another (in A) appears after the occurrence it dominates. Next, perform weighted substitution of B for t_1 in A. In the resulting formula $A]_{t_1}^B$ we denote by $t'_2, \ldots, t'_r$ the occurrences of b appearing instead of $t_2, \ldots, t_r$. More precisely, writing A in the form $UtA_1 \ldots A_hV$, we get $A]_{t_1}^B = UB'V$, where B′ is obtained from B as indicated in 1.4.1 and the occurrences $t_2, \ldots, t_r$ which belong to U and V reappear in UB′V as $t'_2, \ldots, t'_r$. Now perform weighted substitution of B for t'_2 in $A]_{t_1}^B$, and so on.

Example. Letting $A = (aa''a'a)\, b''(a'b)$, perform weighted substitution of $B = a_1c''a'a_1$ for a'. The substitution instance of $a'a$ is then $ac''a'a$ and that of $a'b$ is $bc''a'b$; the final result is $(aa''(ac''a'a))\, b''(bc''a'b)$.

1.4.3. *Simultaneous Weighted Substitution*

Let A be a formula, $b_1, \ldots, b_r$ elements of weights $h_1, \ldots, h_r \geqslant 0$ having one or more occurrences in A, and $B_1, \ldots, B_r$ formulas. Generalizing the preceding definition, we define the *simultaneous substitution instance* $A]^{B_1 \ldots B_r}_{b_1 \ldots b_r}$ as follows. Write A in the form $tA_1 \ldots A_k$. Now, either $k=0$ or $k \geqslant 1$. If $k=0$, then A reduces to t; if t is not an occurrence of any of the $b_1, \ldots, b_r$, we define the substitution result to be t, while if t is an occurrence of b_i the new formula is B_i. If $k \geqslant 1$, assume that the substitution results $A_i]^{B_1 \ldots B_r}_{b_1 \ldots b_r}$ $(i=1, \ldots, k)$ have been obtained. Then if t is not one of the $b_1, \ldots, b_r$, set

$$A]^{B_1 \ldots B_r}_{b_1 \ldots b_r} = t(A_1]^{B_1 \ldots B_r}_{b_1 \ldots b_r}) \ldots (A_k]^{B_1 \ldots B_r}_{b_1 \ldots b_r}).$$

If t is an occurrence of b_i, so that $k=h_i$, then the substitution result is obtained from B_i by simultaneous substitution of the expression $A_j]^{B_1 \ldots B_r}_{b_1 \ldots b_r}$ (in the ordinary sense) for each atom a_j $(j=1, \ldots, h_i)$ in B_i.

Example. Let $A=(aa''a'a)\, b''(a'b)$; consider the simultaneous substitution of $A'=a_1c''a'a_1$ for a' and $A''=a_1c''b'a_2$ for a''. The substitution instance of $a'a$ is then $ac''a'a$, that of $a'b$ is $bc''a'b$, and that of $aa''a'a$ is $ac''b'(ac''a'a)$; the final result is $(ac''b'(ac''a'a))\, b''(bc''a'b)$.

1.4.4. *Superimposed Substitutions*

Let C be a formula and α an element of weight $h \geqslant 0$. Let us perform weighted substitution of some formula A for α, and then, in the result $C]^A_\alpha$, substitute some formula B for α, writing the result as $C]^A_\alpha]^B_\alpha$. Next, let us substitute B for α in A, giving $A]^B_\alpha$, and then substitute $A]^B_\alpha$ for α in C, giving $C\Big]^{(A]^B_\alpha)}_\alpha$. *The two formulas thus obtained are the same*; i.e.,

$$C]^A_\alpha]^B_\alpha = C\Big]^{(A]^B_\alpha)}_\alpha.$$

We prove this in four steps.

(1) *The case* $h=0$. Let us write C in the form $UaVa \ldots aW$, where the words U, V, …, W do not contain a. Then

$$C]^A_a = UAVA \ldots AW \quad \text{and} \quad C]^A_a]^B_a = UA]^B_a VA]^B_a \ldots A]^B_a W,$$

which is precisely $C\Big]^{A]^B_a}_a$.

(2) More generally, let $a_1, \ldots, a_h$ be elements of weight 0 in C. Consider the simultaneous substitution of a formula A_i, and then of a formula B_i,

for each a_i $(i=1,\dots,h)$ in C; the result is $(\mathrm{C}]^{\mathrm{A}_1\dots\mathrm{A}_h}_{a_1\dots a_h})]^{\mathrm{B}_1\dots\mathrm{B}_h}_{a_1\dots a_h}$. Now, for each a_i in C, substitute $\mathrm{A}_i]^{\mathrm{B}_1\dots\mathrm{B}_h}_{a_1\dots a_h}$. *The two formulas obtained are the same*; i.e.,

$$(\mathrm{C}]^{\mathrm{A}_1\dots\mathrm{A}_h}_{a_1\dots a_h})]^{\mathrm{B}_1\dots\mathrm{B}_h}_{a_1\dots a_h}=\mathrm{C}\Bigg]^{(\mathrm{A}_1]^{\mathrm{B}_1\dots\mathrm{B}_h}_{a_1\dots a_h})\dots(\mathrm{A}_h]^{\mathrm{B}_1\dots\mathrm{B}_h}_{a_1\dots a_h})}_{a_1\qquad\dots\qquad a_h}.$$

To see this write C as $\mathrm{U}\,a_i\,\mathrm{V}\,a_j\dots a_k\,\mathrm{W}$, where U, V,..., W do not contain any of the $a_1,\dots,a_h$. Both the above substitutions give

$$\mathrm{U}(\mathrm{A}_i]^{\mathrm{B}_1\dots\mathrm{B}_h}_{a_1\dots a_h})\ \mathrm{V}\ (\mathrm{A}_j]^{\mathrm{B}_1\dots\mathrm{B}_h}_{a_1\dots a_h})\dots(\mathrm{A}_k]^{\mathrm{B}_1\dots\mathrm{B}_h}_{a_1\dots a_h})\ \mathrm{W}.$$

(3) Let a_1 be the atom with index 1 and β an element of weight $\geqslant 1$. Then, given formulas A, B, C, we have the equality

$$\mathrm{C}]^{\mathrm{A}}_{a_1}]^{\mathrm{B}}_{\beta}=\mathrm{C}]^{\mathrm{B}}_{\beta}\Bigg]^{\mathrm{A}]^{\mathrm{B}}_{\beta}}_{a_1}.$$

▷ Let us first assume that C is a single element c. Now, either $c\neq a_1$ or $c=a_1$. If $c\neq a_1$, then the two formulas in the assertion reduce to c. If, on the other hand, $c=a_1$, then $\mathrm{C}]^{\mathrm{A}}_{a_1}]^{\mathrm{B}}_{\beta}=\mathrm{A}]^{\mathrm{B}}_{\beta}$ and $\mathrm{C}]^{\mathrm{B}}_{\beta}=a_1$, so that by substituting $\mathrm{A}]^{\mathrm{B}}_{\beta}$ for a_1 we get $\mathrm{A}]^{\mathrm{B}}_{\beta}$ itself.

Let C have the form $t\mathrm{C}_1\dots\mathrm{C}_h$. Then, since t is of positive weight h, we must have $t\neq a_1$. Now, either $t\neq\beta$ or $t=\beta$. If $t\neq\beta$, then $\mathrm{C}]^{\mathrm{A}}_{a_1}]^{\mathrm{B}}_{\beta}$ reduces to t followed by the formulas $\mathrm{C}_i]^{\mathrm{A}}_{a_1}]^{\mathrm{B}}_{\beta}$ $(i=1,\dots,h)$. Furthermore, $\mathrm{C}]^{\mathrm{B}}_{\beta}\Big]^{\mathrm{A}]^{\mathrm{B}}_{\beta}}_{a_1}$ reduces to t followed by the same formulas if we assume the assertion true for each C_i, so that it is therefore again true for C. But if $t=\beta$, then $\mathrm{C}=\beta\mathrm{C}_1\dots\mathrm{C}_h$ and so

$$\mathrm{C}]^{\mathrm{A}}_{a_1}=\beta(\mathrm{C}_1]^{\mathrm{A}}_{a_1})\dots(\mathrm{C}_h]^{\mathrm{A}}_{a_1})\quad\text{and}\quad \mathrm{C}]^{\mathrm{A}}_{a_1}]^{\mathrm{B}}_{\beta}=\mathrm{B}\Bigg]^{\mathrm{C}_1]^{\mathrm{A}}_{a_1}]^{\mathrm{B}}_{\beta}\dots\mathrm{C}_h]^{\mathrm{A}}_{a_1}]^{\mathrm{B}}_{\beta}}_{a_1\qquad\dots a_h}.$$

On the other hand

$$\mathrm{C}^{\mathrm{B}}_{\beta}=\mathrm{B}\Bigg]^{\mathrm{C}_1]^{\mathrm{B}}_{\beta}\dots\mathrm{C}_h]^{\mathrm{B}}_{\beta}}_{a_1\qquad\dots a_h},$$

so that

$$(\mathrm{C}]^{\mathrm{B}}_{\beta})\Bigg]^{\mathrm{A}]^{\mathrm{B}}_{\beta}}_{a_1}=\left(\mathrm{B}\Bigg]^{\mathrm{C}_1]^{\mathrm{B}}_{\beta}\dots\mathrm{C}_h]^{\mathrm{B}}_{\beta}}_{a_1\qquad\dots a_h}\right)\Bigg]^{\mathrm{A}]^{\mathrm{B}}_{\beta}\ a_2\dots a_h}_{a_1\qquad a_2\dots a_h}.$$

By part 2, the last formula can be obtained from B by substituting $C_i]^B_\beta \Big]_{a_1}^{A]^B_\beta}$ for a_i, $i=1,\ldots,h$. Assuming the assertion to be true for $C_1,\ldots,C_h$, this is equivalent to substitution of $C_i]^A_{a_1}]^B_\beta$ for each a_i, and we get the same formula as before. ◁

(4) More generally, let $a_1,\ldots,a_h$ be atoms, A_i a formula for each i $(i=1,\ldots,h)$, β an element of weight 1. Then, given the formulas B and C, *we have the equality*

$$C]^{A_1\ldots A_h}_{a_1\ldots a_h}]^B_\beta=(C]^B_\beta)\Big]_{a_1\ \ldots a_h}^{A_1]^B_\beta\ldots A_h]^B_\beta}.$$

Proof of 1.4.4. ▷ If $h=0$, this is part 1. Let us assume, therefore, that $h\geqslant 1$ and, first, that C reduces to an element c. Since c is of weight 0, we have $c\neq\alpha$, and the two sides of the equality reduce to c.

Next assume C to have the form $tC_1\ldots C_k$. Now, either $t\neq\alpha$ (this being the case, in particular, when $h\neq k$) or $h=k$ and $t=\alpha$. In the first case, $t\neq\alpha$, $C]^A_\alpha]^B_\alpha$ is t followed by the formulas $C_i]^A_\alpha]^B_\alpha$ $(i=1,\ldots,k)$. Furthermore, $C\Big]_\alpha^{A]^B_\alpha}$ is t followed by the formulas $C_i\Big]_\alpha^{A]^B_\alpha}$. If the assertion is true for each C_i, it is true for C as well. In the second case, $h=k$ and $t=\alpha$, we have $C=\alpha C_1\ldots C_h$, whence $C]^A_\alpha=A\Big]_{a_1\ \ldots\ a_h}^{C_1]^A_\alpha\ldots C_h]^A_\alpha}$ and the left-hand side of the equality becomes $\left(A\Big]_{a_1\ \ldots\ a_h}^{C_1]^A_\alpha\ldots C_h]^A_\alpha}\right)\Big]^B$. On the other hand,

$$C\Big]_\alpha^{A]^B_\alpha}=(A^B_\alpha)\Big]_{a_1\ \ldots a_h}^{C_1]_\alpha^{A]^B_\alpha}\ldots C_h]_\alpha^{A]^B_\alpha}}$$

Since the assertion is assumed true for each C_i $(i=1,\ldots,h)$, we can replace each $C_i\Big]_\alpha^{A]^B_\alpha}$ by $C_i]^A_\alpha]^B_\alpha$ in the last formula. By part 4 with C replaced by A, β by α, and each A_i by $C_i]^A_\alpha$, this formula reduces to $A\Big]_{a_1\ \ldots a_h}^{C_1]^A_\alpha\ldots C_h]^A_\alpha}\Big]_\alpha^B$, and it is thus identical to the formula above. ◁

EXERCISES

1

A set is said to be *finite*, in the sense of [TAR, 1924] (see also [SUP, 1960]), if every non-empty set of subsets B_i of A contains a minimal element under inclusion, in other words, a subset B such that no B_i is a proper subset of B. By passing to complements, one sees that this is true if and only if every nonempty set of subsets of A contains a maximal element under inclusion.

(1) Show that the empty set and every singleton (set consisting of a single element) are finite; show that if A is finite, then every subset of A is finite.

(2) Given two finite sets A and B, show that the union $A \cup B$ is finite; in particular, show that the set obtained by adjoining a single element to A is finite.

(3) Show that the image of a finite set under any mapping is a finite set.

(4) Let A be a finite set and a an element; show that, if the power set of A is finite, then the power set of $A \cup \{a\}$ is finite. Hence prove by induction that if A is finite, the power set of A is finite (consider those subsets X of A such that the power set of X is finite and take an X which is maximal under inclusion). Show that if A and B are finite, the cartesian product of A and B is finite and the set of mappings of A into B is finite.

(5) Show that a finite set is not equipollent to any of its proper subsets (Dedekind's condition). Show that, assuming the Axiom of Choice, one obtains the converse: every infinite set is equipollent to a proper subset.

(6) Show that A is finite if and only if it satisfies the following condition: If for every set $\mathscr{A}$ the empty set $0 \in \mathscr{A}$, and for every $X \in \mathscr{A}$ and $x \in A$ the union $X \cup \{x\} \in \mathscr{A}$, then $A \in \mathscr{A}$ (definition of finite sets due to [SIE, 1918] and [KUR, 1920]). If a proposition is true for the empty set, and whenever it is true for A, it is also true for the set obtained by adjoining an arbitrary element to A, then it is true for any finite set.

(7) Show that two finite sets, or two sets only one of which is finite, are comparable, i.e., there exists a subset of one equipollent to the other.

(8) Show that A is finite if and only if there exists a total ordering, or chain on A, such that every nonempty subset of A contains a minimal and a maximal element; then every total ordering of A satisfies this condition and all the resulting chains are isomorphic ([STA, 1907]; see the definitions in 3.1 and 3.2).

2

We shall employ the set-theoretical axioms mentioned at the beginning of the chapter, including the axiom of regularity: every nonempty set a has an element which is disjoint from a, and its corollary: for all a, $a \notin a$. The axiom of infinity will be mentioned explicitly at each of its rare appearances. Tarski's definition of finite sets (see Exercise 1) will be frequently employed. We define an *ordinal* as a set α satisfying two conditions: $\in$-*transitivity* (for every x and y, if $x \in y$ and $y \in \alpha$, then $x \in \alpha$) and $\in$-*comparability* (for every x and y, if $x \in \alpha$ and $y \in \alpha$, then either $x \in y$, $y \in x$, or $x = y$). *Examples:* the empty set 0, the set $1 = \{0\}$, the set $2 = \{0, 1\}$. Give examples of sets satisfying only one or neither of the two conditions.

(1) Show that, for every nonempty ordinal α, we have $0 \in \alpha$; if α is an ordinal, then $\alpha \cup \{\alpha\}$ is an ordinal (which we shall call the *successor* of α); show that every element of an ordinal is an ordinal.

(2) Show that if α and β are two ordinals, then the condition $\alpha \in \beta$ is equivalent to proper inclusion $\alpha \subset \beta$; this will be denoted by $\alpha < \beta$, and we shall write $\alpha \leqslant \beta$ when $\alpha < \beta$ or $\alpha = \beta$. Show that any ordinal α is the set of the ordinals strictly smaller than α.

(3) Given a set a of ordinals, the intersection of the elements in a is an ordinal belonging

to a, which is thus the minimal element in a with respect to $\leqslant$. Show that, for two ordinals α and β, either $\alpha<\beta$, $\beta<\alpha$, or $\alpha=\beta$.

(4) Given a set a of ordinals, show that the union of the elements of a is an ordinal which does not necessarily belong to a itself and which is the least upper bound of a with respect to $\leqslant$. Show that the set of finite ordinals, provided it exists (axiom of infinity), is not finite. Show that the set of all ordinals does not exist (Burali-Forti paradox).

(5) Repeat the above without employing the axiom of regularity, but with ordinals defined as $\in$-transitive, $\in$-comparable, and $\in$-regular sets in the sense that for every nonempty subset u of the ordinal there exists $x\in u$ such that x is disjoint from u (see [BER, 1960]).

3

A *natural number* is any finite ordinal (in Tarski's sense; see Exercises 1 and 2); recall that the *successor* of an ordinal α is $\alpha\cup\{\alpha\}$.

(1) Show that the successor of a natural number is also a natural number, every element of a natural number is a natural number, and every natural number $\neq 0$ is the successor of a unique natural number, which we shall call its *predecessor*. Show that α is a natural number if and only if it satisfies the following recursive condition: if a set contains 0 and, whenever it contains $x\in\alpha$ also contains $x\cup\{x\}$, then it contains α.

(2) Show that the axiom of regularity and the definition of natural number imply that for no natural number α does there exist a "cycle" of arbitrary sets a, indexed by the natural numbers $\leqslant\alpha$, such that

$$a_0\in a_1\in a_2\in\cdots\in a_\alpha\in a_0.$$

More generally, a sequence indexed by the natural numbers α such that $a_{\alpha+1}\in a_\alpha$ ($\alpha+1=$ the successor of α) is necessarily finite. Show that this last condition implies the axiom of regularity, but the Axiom of Choice is needed to prove this.

4

Let us consider 26 elements, represented by the letters of the alphabet, a, b, c, ..., z, and let a *word* be any finite sequence of letters, without regard to its pronunciation or meaning in any ordinary language, so that "rstzr" and "ztuin" will be words on a par with "rat" or "shovel". We define the following three functions taking words onto words:

(a) the function f inverts the order of the letters: f(are)=era, f(deer)=reed;

(b) the function g interchanges the first two letters (leaving the word unchanged if it consists of only one letter): g(art)=rat, g(altitude)=latitude, $g(a)=a$.

(1) Verify (for the word "team") that neither gf, $(gf)^2$ nor $(gf)^3$ is the identity. Show that, on the other hand, $(gf)^4$ is the identity for every word of 1 letter, 2 letters, 4 or more letters, and that $(gf)^3$ is the identity for every word of 3 letters.

What is the smallest integer p such that $(gf)^p$ is the identity for every word?

(2) Define a function k such that $k^2=f$.

The solution is easy when we are allowed to add a new element α to the letters of the alphabet and to let k act on the words in this enlarged alphabet, while f remains defined on the words in the initial alphabet.

(3) Retaining the original alphabet, show that there is no function k which is a simple permutation of the letters in each word.

(4) Classify words as symmetric (example: level) and asymmetric; group the latter in pairs (each word with its image under f), and enumerate these pairs arbitrarily; now find a function k as required.

(5) What condition must the number of letters of the alphabet satisfy for the existence

of a function k such that $k^2=f$ which preserves the number of letters in each word?

5

Recall that when we start with a formula and substitute a formula for a term of weight 0 the result is a formula.

(1) Prove the following converse: given a formula A = UBV, where B is a formula and U and V are two words, then for every element x of weight 0 the word UxV is a formula (induction on the number of terms in A).

(2) From the existence of a formula X such that UXV is a formula, deduce that for every formula Y the word UYV is a formula. Generalize, replacing X by a sequence of h formulas. Give an example in which X and UXV are formulas, UWV is a formula, but W is not.

(3) If we substitute a formula for the first term in a formula having at least two terms, we do not get a formula. Why?

(4) If in a formula A we remove a term of weight 0 or replace a term of weight 0 by a sequence of several terms of weight 0, we do not get a formula (induction on the number of terms in A).

(5) If $UX_1...X_hV$ is a formula (where $X_1,..., X_h$ are formulas), then, replacing $X_1...X_h$ by a sequence of k formulas, we get a formula if and only if $k=h$.

(6) If in a formula we replace a term of positive weight by a formula, we do not get a formula.

(7) Show that the statement of (4) holds when weight 0 is replaced throughout by $h \geqslant 2$. On the other hand, if we remove a term of weight 1 from a formula or replace it by a sequence of several terms of weight 1, we get a formula.

(8) If we replace one term in a formula by another of different weight, we do not get a formula.

(9) Show that a word U is of zero weight if and only if it can be derived from a formula by permutation of terms (induction on the number of terms). More precisely, for every word U of zero weight and every term t of positive weight in U, there exists a permuted formula of U beginning with t.

Prove the propositions of Section 1.2 in the text, and then the preceding results, using the notion of the weight of a word.

6

Two words U and V are said to be *circularly equivalent* if there exist an initial interval X and a final interval Y in U such that U = XY and V = YX. A set of words circularly equivalent to one of them is called a *circular word.* Since the weight does not change in going from U to V, we shall speak of the *weight* of a circular word.

(1) Show that a circular word of weight 0 contains one, and only one, formula. We can represent a word by a sequence of vectors, each associated with a term and having first component 1 and second component equal to the weight of the term minus 1. The initial point of each vector is the terminal point of the preceding one. Now consider the straight lines with slope $-1/n$ passing through each "corner" of this polygonal line and show that they are all distinct.

(2) Hence show again that a word is of zero weight if and only if it can be obtained from a formula by permutation of terms (Exercise 5.9); furthermore, for every term t of positive weight we can get a formula beginning with t by permutation.

(3) Consider a finite set of elements $a, b,...,$ to each of which is assigned a weight $p(a), p(b),...$ and a positive integer $h(a), h(b),....$ Using factorials, calculate the number of

formulas with $h(a)$ occurrences of a, $h(b)$ occurrences of b, and so on. State a condition under which this number is not zero, and note that the number then depends only on the different integers h and not on the weights.

(4) Applying the preceding to the special case in which there are only two elements, a of weight 0 and b of weight 2, recalculate the number of sequences of h elements a and h elements b, where there are at least as many a's as b's in each final interval (private communication from P. Jullien).

7

Given two words U and V, write $U \leqslant V$ when U is a subsequence of V in the usual sense, i.e., when U is obtained by removing some of the terms in V and renumbering those which remain; for example, art $\leqslant$ warrant.

An ordering (reflexive, transitive, and antisymmetric relation) is called a *partial well-ordering* if every totally ordered restriction has a minimal element (if not empty). A set whose elements are pairwise incomparable is said to be *free*, and the ordering is *finitely free* if every free set is finite. A finitely free partial well-ordering is called a *better-ordering*.

(1) For a finitely free ordering, show that every infinite set of elements contains a totally ordered infinite subset. In a better-ordering one can extract an increasing infinite sequence from any infinite sequence.

(2) Given two orderings A and B, we define their *product* as a relation over ordered pairs of elements, as follows. If x and x' belong to the base* of A and y and y' to that of B, set $(x, y) \leqslant (x', y')$ when $x' \leqslant x'$ in A and $y \leqslant y'$ in B. Show that the product of two better-orderings is a better-ordering.

(3) A is a better-ordering if and only if, for each element x in the base of A, the restriction of A to elements not $\geqslant x$ is also a better-ordering (the same holds with "better-ordering" replaced by "partial well-ordering" or by "finitely free ordering").

(4) Given two sets $\mathscr{U}$ and $\mathscr{V}$ of words, we call the set of concatenations of a word from $\mathscr{U}$ and a word from $\mathscr{V}$ their *concatenation product* $\mathscr{U}\mathscr{V}$. Show that if $\leqslant$ is a better-ordering on $\mathscr{U}$ and on $\mathscr{V}$, it is a better-ordering on their concatenation product $\mathscr{U}\mathscr{V}$. Generalize to the case of an arbitrary finite sequence of sets of words.

(5) Given a fixed set over which all words are considered, and one of these words U, denote by $\breve{U}$ the set of words not $\geqslant U$ and by U' the set consisting of the empty word and the word U. If U and V are two nonempty words, show that $\overline{UV} = \breve{U}\breve{V}$ if the last element in U and the first in V are distinct, and $\overline{UV} = \breve{U}a'\breve{V}$ if a is both the last in U and the first in V.

(6) Show that, *in the set of words over a finite set, the relation $\leqslant$ is a better-ordering* [HIG, 1952]. On the other hand, if we define the relation $U \leqslant V$ as: U is an interval of V, there exist infinite free sets; for example, *aa*, *aba*, *abba*, *abbba*,... (private communication from P. Jullien; for definitions see Chapters 1 and 3).

NOTE

* I.e., the domain of the ordering A. – Trans.

CHAPTER 2

CONNECTIONS AND CONNECTIVE FORMULAS*

2.1. TRUTH VALUES; CONNECTION, ARITY; NEGATION, CONJUNCTION, DISJUNCTION, IMPLICATION, DEDUCIBILITY

The connections originate from the expressions "and", "or", "not", "if...then" of every day language.

The *connective calculus*, also known as the *propositional calculus*, already reached a high level of development in [BOO, 1847] and [MOR, 1847]. A great stcp forward was identification of each connection with a "truth table", that is, a function having two values which we denote by the symbols + and − (we shall also call these values "true" and "false"). This method goes back to [FRE, 1879] (see also [PEI, 1885] and [SCH, 1895]), but had to wait for [LUK, 1921], [POS, 1921] and [WIT, 1922] to receive systematic development.

2.1.1. We shall call a function α an *h-ary connection* if, with each *h-tuple* or sequence of h elements $a_1, \dots, a_h$ each of which has the value + or −, the function α associates a value $\alpha(a_1, \dots, a_h)$ equal to + or −. The integer h is called the *arity* of α, and for $h=1, 2$, or 3 we shall call the connections respectively *unary*, *binary*, or *ternary*. Instead of $a_1, a_2, a_3, \dots$ we shall also use the notation $a, b, c, \dots$.

Since there are 2^h sequences of h values + and −, there must be $2^{(2^h)}$ h-ary connections. In particular, there exist four unary connections: the connection + such that $+(a)$ is always equal to +; the connection −, such that $-(a)$ is always equal to −; the connection ( ), such that $(a)=a$; and finally *negation* $\neg$, where $\neg a$ (read: "not a") takes the value − or + according to whether a equals + or −, respectively.

Of the 16 binary connections we mention the connection $+_2$, with $+_2(a, b)$ always equal to +; the connection $-_2$; *conjunction* $\bigwedge$, where $\bigwedge(a, b)$ (read "a and b") is equal to + if and only if $a=+$ and $b=+$; *disjunction* $\bigvee$, where $\bigvee(a, b)$ (read "a or b") is equal to + if and only if $a=+$ or $b=+$; *implication* $\Rightarrow$, where $\Rightarrow(a, b)$ (read "if a then b" or "a

implies b") is equal to $+$ when either $a=b=+$, or $a=-$ (and b is either $+$ or $-$); and finally *biconditional* [sometimes called *equivalence*] $\Leftrightarrow$, where $\Leftrightarrow(a, b)$ (read "a if and only if b", or "a and b imply each other") is equal to $+$ if and only if $a=b=+$ or $a=b=-$.

Of the h-ary connections for $h\geqslant 3$, we mention $+_h$, which is always equal to $+$, and $-_h$, which is always equal to $-$; we shall often use *h-ary conjunction* $\bigwedge_h$, where $\bigwedge_h(a_1,\ldots, a_h)$ is equal to $+$ if and only if each of the $a_1,\ldots, a_h$ has the value $+$, and *h-ary disjunction* $\bigvee_h$, where $\bigvee_h(a_1,\ldots, a_h)$ is equal to $+$ if and only if at least one of the $a_1,\ldots, a_h$ is equal to $+$.

It will be convenient to regard the truth values $+$ and $-$ as (the only) two 0-ary connections.

2.1.2. Let α and β be two h-ary connections; we shall say that *β is deducible* from α, writing $\alpha\vdash\beta$, if every h-tuple $a_1,\ldots, a_h$ *of* $+$'s and $-$'s which gives α the value $+$ also gives β the value $+$.

Example. $\bigwedge\vdash\bigvee$, since whenever $a\bigwedge b=+$ we have $a=b=+$, and therefore $a\bigvee b=+$. We also have $\bigwedge\vdash\Rightarrow$, since whenever $a\bigwedge b=+$ we have $a\Rightarrow b=+$.

More generally, let α and β be two connections of arities h and k; set $\alpha\vdash\beta$ if every h-tuple which gives α the value $+$, reduced to its first k terms if $k\leqslant h$ or expanded by $k-h$ arbitrary values $+$ or $-$ if $k>h$, also gives β the value $+$.

Example. $\neg\vdash\Rightarrow$, since whenever $\neg a=+$ we must have $a=-$, which means that $a\Rightarrow b=+$ will hold whether $b=+$ or $b=-$.

The deducibility relation $\vdash$ is an ordering over the h-ary connections for given h (it is reflexive, transitive and antisymmetric). Over connections of arbitrary arity, $\vdash$ is only a pre-ordering (reflexive and transitive relation) [also known as a *quasi-ordering.* – Trans.].

The connections $+_h$ (of arbitrary arity h) are deducible from any connection; any connection is deducible from $-_h$.

If $\alpha\vdash\beta$ and $\beta\vdash\alpha$, we shall say that α and β are *equideducible*, and write $\alpha\dashv\vdash\beta$. Equideducibility is an equivalence relation. If α and β have the same arity and are equideducible, then $\alpha=\beta$. If α is h-ary and β is k-ary $(k>h)$ and equideducible from α, then the value of β is independent of the values $a_{h+1},\ldots, a_k$. It follows that, for each $k\geqslant h$, there exists *one and only one k-ary connection equideducible from α.*

Example. The connections $+_h$ of arities $h=1, 2, 3,\ldots$ always equal to $+$, are equideducible; the same holds for the connections $-_h$. For the unary connection $\neg$, the equideducible connection is $\alpha(a, b)$ which takes the value opposite to a and is independent of b.

2.1.3. THE ORDERING $\vdash$ ON THE h-ARY CONNECTIONS (h-fixed) INDUCES A BOOLEAN ALGEBRA (i.e., a distributive and complemented lattice [with minimal and maximal elements]; these concepts, though classical, will be defined in 7.4.2). With each h-ary connection α, we associate the set of sequences of values $a_1,\ldots, a_h$ which give α the value $+$. Then $\vdash$ is simply the inclusion relation for these sets.

With any two h-ary connections α and β, we associate the following two connections: (1) the *infimum* of α and β, denoted $\bigwedge_{\alpha,\beta}$, equal to $+$ for sequences of values $a_1,\ldots, a_h$ for which $\alpha(a_1,\ldots, a_h)$ and $\beta(a_1,\ldots, a_h)$ both take the value $+$; (2) the *supremum* of α and β, denoted $\bigvee_{\alpha,\beta}$, equal to $+$ when either $\alpha(a_1,\ldots, a_h)=+$ or $\beta(a_1,\ldots, a_h)=+$. These correspond to the intersection and union of sets of sequences. In other words, if ξ is some h-ary connection, then $\xi\vdash\alpha$ and $\xi\vdash\beta$ if and only if $\xi\vdash\bigwedge_{\alpha,\beta}$; and $\alpha\vdash\xi$ and $\beta\vdash\xi$ if and only if $\bigvee_{\alpha,\beta}\vdash\xi$. The operation which associates $\bigwedge_{\alpha,\beta}$ with α and β is commutative and associative; the same holds for $\bigvee_{\alpha,\beta}$.

Each of the operations $\bigwedge_{\alpha,\beta}$ and $\bigvee_{\alpha,\beta}$ is distributive over the other: i.e.

$$\bigvee_{(\wedge_{\alpha,\beta}),\gamma}=\bigwedge_{(\vee_{\alpha,\gamma},\vee_{\beta,\gamma})}$$

and similarly for $\bigvee$ and $\bigwedge$ interchanged. This is none other than the classical distributivity of union and intersection.

Finally, the set of h-ary connections with the ordering $\vdash$ and the operations $\bigwedge_{\alpha,\beta}$ and $\bigvee_{\alpha,\beta}$, is complemented; the minimal element is the connection $-_h$, the maximal element is the connection $+_h$. If with each h-ary connection α we associate its "relative negation" $\neg_\alpha$ which always takes the opposite value to α, we see that $\bigwedge_{\alpha,\neg_\alpha}$ is the connection $-_h$ and $\bigvee_{\alpha,\neg_\alpha}$ is the connection $+_h$.

There is a "cubic" representation of the Boolean algebra of h-ary connections. Consider the 2^h sequences of h values $+$ or $-$; with each of these, say s_i $(i=1,\ldots, 2^h)$, associate the connection α_i equal to $+$ only for the sequence s_i. Represent these connections α_i by 2^h independent vectors. Next, with the operation $\bigvee_{\alpha,\beta}$ on the connections associate addition of vectors. The zero vector will thus correspond to the connection $-_h$; the connection α taking the value $+$ for the sequences $s_i, s_j, \ldots$ is associated with the resultant (sum) of the vectors $\alpha_i, \alpha_j, \ldots$. The end-points of the vectors are $2^{(2^h)}$ in number, arranged at the vertices of a parallelepiped (or cube) of dimension 2^h. The case shown here is $h=2$, with $2^{(2^2)}=16$ binary connectives; the 4-dimensional cube is represented by two 3-dimensional cubes translated along the vector shown beneath them.

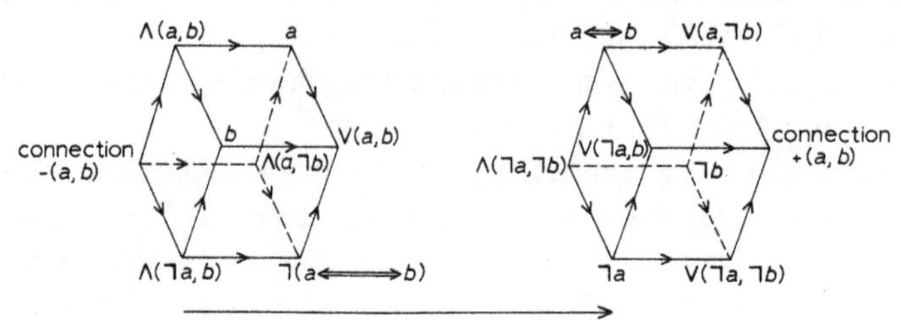

2.1.4. A set with a pre-ordering (reflexive, transitive, not necessarily antisymmetric relation), denoted $\vdash$, is said to be a *pre-lattice* if every pair α, β of elements has an infimum $\bigwedge_{\alpha,\beta}$ and a supremum $\bigvee_{\alpha,\beta}$ satisfying the conditions: (1) $\xi \vdash \alpha$ and $\xi \vdash \beta$ is equivalent to $\xi \vdash \bigwedge_{\alpha,\beta}$, and (2) $\alpha \vdash \xi$ and $\beta \vdash \xi$ is equivalent to $\bigvee_{\alpha,\beta} \vdash \xi$. There may be several elements satisfying these conditions, but if so they are all equivalent, in the sense that each precedes the other in the ordering $\vdash$; we shall denote this equivalence by $-\vdash$. We get a lattice by considering the ordering induced by $\vdash$ on the equivalence classes with respect to $-\vdash$. We can now obtain the obvious notion of a *distributive* pre-lattice, by simply replacing the distributivity equalities by $-\vdash$-equivalences. A distributive pre-lattice is *Boolean* if it possesses minimal elements $-$ (all equivalent), maximal elements $+$, and, for each α, an element $\neg_\alpha$ such that $\bigwedge_{\alpha, \neg_\alpha}$ is one of the $-$ and $\bigvee_{\alpha, \neg_\alpha}$ is one of the $+$. We get a distributive lattice and a Boolean algebra by passing to equivalence classes.

The *pre-ordering of deducibility* $\vdash$ makes the set of connections of arbitrary arity a *Boolean pre-lattice.* If α is h-ary and β is k-ary ($k>h$), we let $\bigwedge_{\alpha,\beta}$ be the k-ary connection which is equal to $+$ when both $\alpha(a_1, \ldots, a_h)=+$ and $\beta(a_1, \ldots, a_h, a_{h+1}, \ldots, a_k)=+$. We define $\bigvee_{\alpha,\beta}$ similarly. Finally, $\neg_\alpha$ is defined to be the h-ary connection taking the value opposite to α.

2.2. Connective formula; value, tautology, contradiction, deducibility

Connections can obviously be combined, since their domains of definition and ranges both consist of the values $+$ and $-$. For example, it is clear that $\Leftrightarrow (a, b)$ can be obtained from $\Rightarrow$ and $\bigwedge$ as $\bigwedge(\Rightarrow(a, b), \Rightarrow(b, a))$, and $\Rightarrow(a, b)$ is simply $\bigvee(\neg a, b)$. Conversely, given a connection such as $\Rightarrow(a, b)$ one can define new connections by changing the order of the arguments ($\Rightarrow(b, a)$) or identifying them ($\Rightarrow(a, a)$).

The systematic study of these operations on connections is based on the notion of *connective formula*. Associate with each positive integer i an element, called an *atom* [or *individual*], denoted a_i (with the convention that $a_1, a_2, a_3, \ldots$ will also be written as $a, b, c, \ldots$). A formula is called a connective formula if its terms are:

(1) atoms, assigned weight 0;

(2) connections, each assigned weight equal to its arity.

A connective formula whose connections have been chosen from $\alpha, \beta, \gamma, \ldots$ will be called a *formula in* $\alpha, \beta, \gamma, \ldots$. For example, $a_1 \Rightarrow (a_2 \Rightarrow a_1)$ is a formula in $\Rightarrow$, while $a_1 \bigwedge (a_2 \bigvee a_1)$ is a formula in $\bigwedge, \bigvee$.

2.2.1. Let A be a connective formula, and let us assign to each atom in A one of the values $+$ or $-$; such a sequence of values will be called a system of values *assignable to* A. Each such system *gives* A a value $+$ or $-$ (or, equivalently, A *takes* a value $+$ or $-$) defined as follows:

If A is an atom a_i, the value of A is that assigned to a_i.

If A is of the form $\alpha A_1 \ldots A_h$ and $A_1, \ldots, A_h$ take the values $u_1, \ldots, u_h$ (each of which is $+$ or to $-$), then the value taken by A is $\alpha(u_1, \ldots, u_h)$.

Example. Consider the formula $(a_1 \Rightarrow a_2) \bigwedge (a_2 \Rightarrow a_3)$; let the system of values consist of $+$ for a_1, $-$ for a_2, and $+$ for a_3. Then $a_1 \Rightarrow a_2$ takes the value $-$, $a_2 \Rightarrow a_3$ takes the value $+$, and so the whole formula takes the value $-$.

Let h be the maximum of the indices of atoms in A; then the *connection represented by* A is the h-ary connection $\alpha(a_1, \ldots, a_h)$ which takes the same value as A for all h-tuples of $+$ and $-$ assigned to the atoms in A.

Examples. Let α and β be of the same arity h. Then the formula $\bigwedge \alpha a_1 \ldots a_h \beta a_1 \ldots a_h$, also written as $(\alpha a_1 \ldots a_h) \bigwedge (\beta a_1 \ldots a_h)$, represents the connection $\bigwedge_{\alpha, \beta}$ of 2.1.4. Similarly, $\neg \alpha a_1 \ldots a_h$ represents the connection $\neg_\alpha$.

2.2.2. A connective formula A which takes the value + for every system assignable to it is called a *tautology*; we write $\vdash$A. We use the term *contradiction* for a formula A which takes the value $-$ for every system, writing A$\vdash$.

Examples.

$$\vdash a \vee \neg a, \qquad \vdash a \Rightarrow a;$$
$$\vdash ((a \Rightarrow b) \wedge (b \Rightarrow c)) \Rightarrow (a \Rightarrow c);$$
$$a \wedge \neg a \vdash, \qquad a \Leftrightarrow \neg a \vdash.$$

If we replace all the occurrences of an atom in a tautology by the same formula, we get a tautology; the same holds for contradictions.

Let A and B be two formulas; we say that B *is deducible from* A, writing A$\vdash$B, if the connection represented by B is deducible from that represented by A. We can also give a direct definition, starting from a system of values (+ or $-$) assigned to the atoms in A, which gives A the value +; this we complete by assigning arbitrary values to the atoms in B which do not appear in A, suppressing the values of those atoms in A which do not appear in B. If the system thus modified, which is assignable to B, always gives B the value +, then B is deducible from A.

Examples.

$$a \wedge b \vdash a \vdash a \vee b;$$
$$(a \Rightarrow b) \wedge (b \Rightarrow c) \vdash a \Rightarrow c;$$
$$a \wedge (b \vee \neg b) \vdash a.$$

The deducibility relation $\vdash$ on connective formulas is a pre-ordering (reflexive and transitive relation); it is not antisymmetric, for a and $a \wedge (b \vee \neg b)$, which do not have the same atoms, or a and $a \wedge (a \vee \neg a)$, which have the same atom but are distinct, are nevertheless deducible from each other.

A *is a tautology if and only if it is deducible from any formula;* A *is a contradiction if and only if any formula is deducible from* A.

Deducibility and the connection $\Rightarrow$ (if ... then) are related by the following theorem:

A$\vdash$B *if and only if the formula* A$\Rightarrow$B *is a tautology.*

If two formulas A and B are deducible from each other, they are said to be *equideducible*, and we write A$-\vdash$B. This is an equivalence relation.

Examples.

$$\neg\neg a \dashv\vdash a;$$
$$a \Rightarrow b \dashv\vdash \neg a \vee b$$

(one might say that $\Rightarrow$ is thus expressed in terms of $\neg$ and $\vee$)

$$a \vee b \dashv\vdash (\neg a) \Rightarrow b \dashv\vdash (\neg b) \Rightarrow a$$

($\vee$ expressed in terms of $\neg$ and $\Rightarrow$)

$$\neg(a \wedge b) \dashv\vdash \neg a \vee \neg b,$$

and consequently

$$a \wedge b \dashv\vdash \neg(\neg a \vee \neg b)$$

($\wedge$ expressed in terms of $\neg$ and $\vee$);

$$a \vee b \dashv\vdash \neg(\neg a \wedge \neg b)$$

($\vee$ expressed in terms of $\neg$ and $\wedge$);

$$(a \wedge b) \vee c \dashv\vdash (a \vee c) \wedge (b \vee c)$$

(distributivity of $\vee$ over $\wedge$; an analogous equivalence is obtained by interchanging $\wedge$ and $\vee$).

Connections represented by two equideducible formulas are themselves equideducible.

Example. $\neg a$ and $\neg a \wedge (b \vee \neg b)$, representing the connection $\neg$ and the connection $\alpha(a, b)$ equideducible from $\neg$, depend only on a and take the value of $\neg a$.

In particular, if A and B have the same atoms, or the same atom has maximum index in both, and they are equideducible, then they represent the same connection; for example,

$$a \quad \text{and} \quad \neg a \wedge (a \vee \neg a).$$

$A \dashv\vdash B$ if and only if the formula $A \Leftrightarrow B$ is a tautology.

Deducibility defines an ordering (reflexive, transitive and antisymmetric relation) on the classes of equideducible formulas; the minimal element is the class of contradictions and the maximal element the class of tautologies.

2.2.3. With each h-ary connection α we associate an *operation* α which transforms every sequence of h connective formulas $A_1, \ldots, A_h$ into the formula $\alpha A_1 \ldots A_h$.

These operations are compatible with equideducibility; for if

$$A_1 \dashv\vdash A'_1, \ldots, A_h \dashv\vdash A'_h, \quad \text{then} \quad \alpha A_1 \ldots A_h \dashv\vdash \alpha A'_1 \ldots A'_h.$$

We shall therefore associate with α, in an obvious manner, an *operation* α *on the classes of equideducible formulas.*

Note that the operation $\neg$, on either formulas or classes, changes the direction of the deduction: if $A \vdash B$, then $\neg B \vdash \neg A$.

The operations $\bigwedge$, $\bigvee$ and $\neg$ make the set of formulas a Boolean pre-lattice with respect to deducibility, and the classes a Boolean algebra (see 2.1.3 and 2.1.4).

2.3. Formula in $\neg$, $\bigwedge$, $\bigvee$; dual; conjunctive and disjunctive formulas

2.3.1. *Let* A *be a connective formula in* $\neg$, $\bigwedge$, $\bigvee$; *let* A′ *be the formula obtained from* A *by replacing each occurrence of an atom* a_i *by* $\neg a_i$ *and then interchanging* $\bigwedge$ *and* $\bigvee$ *at each of their occurrences; then* $A' \dashv\vdash \neg A$.

▷ The statement is obvious if A reduces to a single symbol, which must necessarily be an atom a_i; for then $A' = \neg a_i \dashv\vdash \neg A$.

Let us assume the statement true for formulas of less than n terms and show that it holds for a formula of n terms ($n \geqslant 2$). Now, A has one of the forms: (1) $\neg B$, (2) $B_1 \bigvee B_2$, or (3) $B_1 \bigwedge B_2$. If A is $\neg B$, then, letting B′ denote the formula obtained from B by the transformation indicated in the statement, we get $A' = \neg B'$, and consequently

$$A' = \neg B' \dashv\vdash B \dashv\vdash \neg A.$$

If A is $B_1 \bigvee B_2$, then, letting B'_1 and B'_2 denote the transforms of B_1 and B_2 as indicated, we get $A' = B'_1 \bigwedge B'_2$; since B_1 and B_2 contain less than n terms, we have $B'_1 \dashv\vdash \neg B_1$ and $B'_2 \dashv\vdash \neg B_2$; consequently

$$A' = B'_1 \bigwedge B'_2 \dashv\vdash \neg B_1 \bigwedge \neg B_2 \dashv\vdash \neg(B_1 \bigvee B_2) = \neg A.$$

Similar reasoning holds if A is $B_1 \bigwedge B_2$. ◁

2.3.2. Given a formula A in $\neg$, $\bigwedge$, $\bigvee$, the *dual of* A is the formula A* obtained by interchanging the connections $\bigwedge$ and $\bigvee$ at all their occurrences. Clearly,

$$(A^*)^* = A, \qquad (\neg A)^* = \neg(A^*). \qquad (A \bigwedge B)^* = A^* \bigvee B^*,$$

and conversely, with $\bigwedge$ and $\bigvee$ interchanged. Furthermore, we have the following *duality lemmas*:

(1) *If* A *is a tautology, then* A* *is a contradiction; if* A *is a contradiction, then* A* *is a tautology.*

(2) *If* $A \vdash B$, *then* $B^* \vdash A^*$.

▷ (1) The formula A′ in 2.3.1 is obtained from A by interchanging $\bigvee$ and $\bigwedge$ and then replacing each occurrence of an atom a_i by $\neg a_i$. We have just seen that $A' \dashv\vdash \neg A$; therefore, if A is a tautology, then A′ is a contradiction: every system of values (+ and −) assigned to the atoms will give A′ the value −. But assigning a system of values to the atoms of A′ amounts to assigning the opposite system of values to the atoms of A*, so that every system of values also gives A* the value −: we have thus established that A* is a contradiction. If A is a contradiction, then $\neg A$ is a tautology, its dual $(\neg A)^* = \neg(A^*)$ is a contradiction, and thus A* is a tautology.

(2) If $A \vdash B$, then $A \Rightarrow B \dashv\vdash \neg A \bigvee B$ is a tautology. The dual of this last formula is $\neg(A^*) \bigwedge (B^*)$, which is a contradiction. Therefore,

$$\neg(\neg A^* \bigwedge B^*) \dashv\vdash A^* \bigvee \neg B^* \dashv\vdash B^* \Rightarrow A^*$$

is a tautology, whence $B^* \vdash A^*$. ◁

2.3.3. Recall that $\bigvee_n$ and $\bigwedge_n$ denote *n*-ary disjunction and *n*-ary conjunction, respectively (see 2.1.1).

A formula A is said to *disjunctive* [BOO, 1847] if it is of the form $\bigvee_n(A_1, \ldots, A_n)$, where each A_i $(1 \leqslant i \leqslant n)$ is of the form $\bigwedge_{n_i}(A_{i,1}, \ldots, A_{i,n_i})$ and each $A_{i,j}$ $(1 \leqslant j \leqslant n_i)$ is either an atom or an atom preceded by negation $\neg$. In other words, A is a disjunction of conjunctions of atoms and negations of atoms.

Similarly, we define a *conjunctive* formula $\bigwedge_n(A_1, \ldots, A_n)$, where each A_i is of the form $\bigvee_{n_i}(A_{i,1}, \ldots, A_{i,n_i})$, and each $A_{i,j}$ is either an atom or an atom preceded by $\neg$.

If a formula A *is not a contradiction, there exists a disjunctive formula which represents the same connection and is therefore equideducible from* A.

Similarly, *if* A *is not a tautology, there exists a conjunctive formula representing the same connection as* A [POS, 1921].

▷ Let A be the formula in question and $a_1, \ldots, a_n$ atoms (the following argument is not affected if the indices are arbitrary integers instead of $1, \ldots, n$); let $s_1, \ldots, s_p$ be the sequences of n values $+$ or $-$ which, when assigned to $a_1, \ldots, a_n$, give the formula the value $+$ (there are of course $1 \leqslant p \leqslant 2^n$ of these). For each s_i, take the formula $A_i = \bigwedge_n(A_{i,1}, \ldots, A_{i,n})$, where $A_{i,j} = a_j$ or $\neg a_j$ according to whether the j-th value of s_i is $+$ or $-$; it is clear that the formula A_i takes the value $+$ for the system s_i and only for that system. The formula $\bigvee_p(A_1, \ldots, A_p)$ is thus equideducible from A and satisfies the conditions of the first assertion. The second is proved by interchanging $\bigwedge$ and $\bigvee$.

Note that in either case the atoms of the conjunctive or disjunctive formula obtained are $a_1, \ldots, a_n$, where n is the maximum index of atoms in A; the formula thus represents the same connection as A.◁

For every formula A *there exist a formula in* $\bigwedge, \neg$ *and a formula in* $\bigvee, \neg$ *which are equideducible from* A.

Indeed, in case A is not a tautology, take the equideducible conjunctive formula, i.e., $\bigwedge_p(A_1, \ldots, A_p)$, then the equideducible formula $\neg\bigvee_p(\neg A_1, \ldots, \neg A_p)$, and finally replace the $\bigvee_p$ and $\bigvee_n$ (appearing in $A_1, \ldots, A_p$) in the obvious way by binary disjunctions $\bigvee$. If A is a tautology, take the equideducible formula $a_1 \bigvee \neg a_1$. In either case we get a formula in $\bigvee, \neg$. Similar reasoning yields a formula in $\bigwedge, \neg$.

2.3.4. Given an h-ary connection α, we define the *dual* of α as the h-ary connection α^* which takes the value $+$ for all systems of h values opposite to the values in the systems which give α the value $-$. In other words, α^* is represented by the formula $\neg\alpha\neg a_1 \ldots \neg a_h$. Obviously, $(\alpha^*)^*$ is identical to α. For example, $+$ and $-$, $\bigwedge$ and $\bigvee$, $a \Leftrightarrow b$ and $a \Leftrightarrow \neg b$ are pairs of duals. If $\alpha \vdash \beta$, then $\beta^* \vdash \alpha^*$.

A connection is said to be *self-dual* if it is identical to its dual. *Examples:* the connection ( ), represented by the atom a; negation $\neg$; the ternary connection represented by $(a \bigwedge b) \bigvee (b \bigwedge c) \bigvee (c \bigwedge a)$ or that represented by $(a \Leftrightarrow b) \Leftrightarrow c$.

If we replace each connection in a formula A by its dual, we get a formula A* which is called the *dual* of A; this gives the same definition as in 2.3.2 when A is a formula in $\neg$, $\bigwedge$ and $\bigvee$; the assertions in 2.3.2 remain valid. Connections represented by two dual formulas are duals. If all the connections in a formula are self-dual, then the connection represented by the formula is self-dual.

2.4. COMPLETENESS LEMMA FOR TAUTOLOGIES [POS, 1921]

2.4.1. *Tautologies in* $\neg$, $\bigvee$.

Denote by $\mathscr{A}$ the smallest set of formulas in $\neg$, $\bigvee$ satisfying the following conditions:

(1) for every formula A, $A \bigvee \neg A \in \mathscr{A}$;
(2) if $A \in \mathscr{A}$, then for all B, $A \bigvee B \in \mathscr{A}$;
(3) if $A \in \mathscr{A}$, then $\neg\neg A \in \mathscr{A}$;
(4) if $A \bigvee B \in \mathscr{A}$, then $A \bigvee \neg\neg B \in \mathscr{A}$;
(5) if $A \bigvee B \in \mathscr{A}$, then $B \bigvee A \in \mathscr{A}$;
(6) if $(A \bigvee B) \bigvee C \in \mathscr{A}$, then $(B \bigvee A) \bigvee C \in \mathscr{A}$;
(7) if $(A \bigvee B) \bigvee C \in \mathscr{A}$, then $A \bigvee (B \bigvee C) \in \mathscr{A}$, and conversely;
(8) if $((A \bigvee B) \bigvee C) \bigvee D \in \mathscr{A}$, then $(A \bigvee (B \bigvee C)) \bigvee D \in \mathscr{A}$, and conversely;
(9) if $\neg A \in \mathscr{A}$ and $\neg B \in \mathscr{A}$, then $\neg(A \bigvee B) \in \mathscr{A}$;
(10) if $A \bigvee \neg B \in \mathscr{A}$ and $A \bigvee \neg C \in \mathscr{A}$, then $A \bigvee \neg(B \bigvee C) \in \mathscr{A}$.

The completeness lemma may then be formulated as follows:

$\mathscr{A}$ is the set of all the tautologies in $\neg$, $\bigvee$.

▷ The proof follows [KAL, 1935]. First, we see that *every formula in $\mathscr{A}$ is a tautology*. For example, as far as condition (10) is concerned, let $A \bigvee \neg B$ and $A \bigvee \neg C$ be two tautologies. Then, for a given system of values u, either $A(u) = +$ and so $A \bigvee \neg(B \bigvee C)$ takes the value $+$ for u, or $A(u) = -$; in the latter case,

$$(\neg B)(u) = (\neg C)(u) = +, \quad \text{whence} \quad B(u) = C(u) = -;$$

thus,

$$(B \bigvee C)(u) = -, \quad \text{whence} \quad \neg(B \bigvee C) \quad \text{and} \quad A \bigvee \neg(B \bigvee C)$$

both take the value $+$.

Conversely, let us prove that every tautology in $\neg$, $\bigvee$ belongs to $\mathscr{A}$. Now, no formula with 1, 2 or 3 terms is a tautology; therefore the above assertion is true for tautologies having $\leqslant 3$ terms. Let us assume it to be true for formulas having less than n terms, and prove it for a tautology A having exactly n terms.

First case. A is of the form $\neg A'$. Then A' cannot be an atom a_i, for if so, when a_i takes the value $+$, the formula $A = \neg a_i$ takes the value $-$. Thus, A' is either of the form $\neg A''$ or of the form $A'' \vee A'''$. If it is $\neg A''$, this brings us back to the tautology A'', which, having less than n terms, belongs to $\mathscr{A}$ by the induction hypothesis. Then $A = \neg\neg A'' \in \mathscr{A}$ by (3). If A' is $A'' \vee A'''$, then A'' and A''' must always take the value $-$, whence $\neg A''$ and $\neg A'''$ are tautologies with less than n terms, and so belong to $\mathscr{A}$. Then $A = \neg(A'' \vee A''') \in \mathscr{A}$ by (9).

Second case. A is of the form $A' \vee A''$. Let us expand A' and A'' as disjunctions until we obtain formulas $A_1, \ldots, A_h$, each of which is either an atom or begins with $\neg$. Then $A \in \mathscr{A}$ if and only if $B = (((A_1 \vee A_2) \vee \vee A_3) \vee \ldots) \vee A_h \in \mathscr{A}$. Indeed, we can go from A to B by using conditions (5), (6), (7) and (8), without changing the number of terms. Now, there are three possible cases: (i) $A_1, \ldots, A_h$ are all atoms or atoms preceded by $\neg$; (ii) one of $A_1, \ldots, A_h$ is of the form $\neg\neg A'$; (iii) one of them is of the form $\neg(A' \vee A'')$. In case (i), since formula B is a tautology, two formules – say A_1 and A_2 – are of the form a_i and $\neg a_i$. Thus $B = ((a_i \vee \neg a_i) \vee A_3) \vee \ldots$, whence $B \in \mathscr{A}$ by (1) and (2). In case (ii), i.e., one of the $A_1, \ldots, A_h$ is of the form $\neg\neg A'$, we may assume that this formula is A_h, so that $B = ((A_1 \vee A_2) \vee \ldots) \vee \neg\neg A'$; by condition (4), this brings us back to $((A_1 \vee A_2) \vee \ldots) \vee A'$, which contains less than n terms. In case (iii), i.e., one of the formulas under consideration is of the form $\neg(A' \vee A'')$, the formula B is, say, $((A_1 \vee A_2) \vee \ldots) \vee \neg(A' \vee A'')$; by condition (10), this brings us back to $((A_1 \vee A_2) \vee \ldots) \vee \neg A'$ and $((A_1 \vee A_2) \vee \ldots) \vee \neg A''$, each of which contains less than n terms. $\lhd$

2.4.2. *Variants*

Many analogous sets of conditions are known which still define the same set $\mathscr{A}$ of tautologies in $\neg, \vee$. Many authors have gone to great lengths to obtain conditions which are few in number and simple, even at the expense of rendering the proof of completeness long and difficult.

The conditions of [RUS, 1908] *and* [RUS-WHI, 1910] *as simplified by* [BER, 1926]:

(1) If $A \in \mathscr{A}$ and $\neg A \vee B \in \mathscr{A}$, then $B \in \mathscr{A}$; this condition, called the *rule of detachment in* $\neg, \vee$, is employed by almost all authors;

(2) $\neg(A\bigvee A)\bigvee A \in \mathscr{A}$;
(3) $\neg A\bigvee(B\bigvee A) \in \mathscr{A}$;
(4) $\neg(A\bigvee B)\bigvee(B\bigvee A) \in \mathscr{A}$;
(5) $\neg(\neg A\bigvee B)\bigvee(\neg(C\bigvee A)\bigvee(C\bigvee B)) \in \mathscr{A}$.

The conditions of [NIC, 1917], these are (1), (2) and (5), plus:

(6) $\neg A\bigvee(A\bigvee B) \in \mathscr{A}$;
(7) $\neg(A\bigvee(B\bigvee C))\bigvee(B\bigvee(A\bigvee C)) \in \mathscr{A}$.

The conditions of [GOT, 1947] *and* [RAS, 1949]; these are (1), (2) and (6), plus:

(8) $\neg(\neg A\bigvee B)\bigvee(\neg(C\bigvee A)\bigvee(B\bigvee C)) \in \mathscr{A}$.

For these three sets of conditions see also [CHU, 1956, p. 137].

2.4.3. *Tautologies in* $\neg$, $\bigwedge$

The method is analogous to that of 2.4.1 for $\neg$, $\bigvee$ (see Exercise 5).

Let us denote by $\mathscr{B}$ the set of tautologies in $\neg$, $\bigwedge$; the following conditions are given by [ROSS, 1953, Chapter 4]. $\mathscr{B}$ is the smallest set of formulas in $\neg$, $\bigwedge$ such that:

(1) If $A \in \mathscr{B}$ and $\neg(A\bigwedge\neg B) \in \mathscr{B}$, then $B \in \mathscr{B}$ (rule of detachment in $\neg$, $\bigwedge$);
(2) $\neg(A\bigwedge\neg(A\bigwedge A)) \in \mathscr{B}$;
(3) $\neg((A\bigwedge B)\bigwedge\neg A) \in \mathscr{B}$;
(4) $\neg(\neg(A\bigwedge\neg B)\bigwedge\neg\neg(\neg(B\bigwedge C)\bigwedge\neg\neg(C\bigwedge A))) \in \mathscr{B}$.

Another set of conditions, which includes (1), is given in [CUR, 1952, p. 114], and another in [SOB, 1954, Section 2].

2.4.4. *Tautologies in* $\neg$, $\Rightarrow$

The method is again analogous to that of 2.4.1 (see Exercise 7).

Let us denote the set of tautologies in $\neg$, $\Rightarrow$ by $\mathscr{C}$; the conditions, which date back to [FRE, 1879], were obtained in the following simplified form by (ŁUK-TAR, 1930]. $\mathscr{C}$ is the smallest set of formulas in $\neg$, $\Rightarrow$ such that:

(1) If $A \in \mathscr{C}$ and $A\Rightarrow B \in \mathscr{C}$, then $B \in \mathscr{C}$ (rule of detachment in $\Rightarrow$);
(2) $A\Rightarrow(B\Rightarrow A) \in \mathscr{C}$;
(3) $A\Rightarrow(B\Rightarrow C))\Rightarrow((A\Rightarrow B)\Rightarrow(A\Rightarrow C)) \in \mathscr{C}$;
(4) $(A\Rightarrow B)\Rightarrow(\neg B\Rightarrow\neg A) \in \mathscr{C}$;
(5) $A\Rightarrow\neg\neg A \in \mathscr{C}$;
(6) $\neg\neg A\Rightarrow A \in \mathscr{C}$.

Other conditions of [ŁUK-TAR, 1930]; these are (1), (2) and (3), plus:

(7) $(\neg A \Rightarrow \neg B) \Rightarrow (B \Rightarrow A) \in \mathscr{C}$.

The conditions of [ŁUK, 1929]; these are (1), plus:

(8) $A \Rightarrow (\neg A \Rightarrow B) \in \mathscr{C}$;

(9) $(\neg A \Rightarrow A) \Rightarrow A \in \mathscr{C}$;

(10) $(A \Rightarrow B) \Rightarrow ((B \Rightarrow C) \Rightarrow (A \Rightarrow C)) \in \mathscr{C}$.

For these three sets of conditions see also [CHU, 1956, pp. 119, 129, 160 and 167].

From conditions (1), (2) and (3) we can deduce (10); it thus follows that conditions (1), (2), (3), (8) and (9) alone define $\mathscr{C}$ (see Exercise 8 and [POR, 1958]).

The conditions of [TAR, 1930] *as simplified by* [SUS, 1957]; these are (1), (2) and (3), plus:

(11) $\neg A \Rightarrow (A \Rightarrow B) \in \mathscr{C}$;

(12) $(\neg A \Rightarrow B) \Rightarrow ((A \Rightarrow B) \Rightarrow B) \in \mathscr{C}$.

The conditions of [CUR, 1952, Chapter 2, Theorem 5]; these are (1), (2), (3), (4) and (6), plus:

(13) $(A \Rightarrow \neg A) \Rightarrow \neg A \in \mathscr{C}$.

2.4.5. *Tautologies in* $\Rightarrow$

Let us denote the set of tautologies in $\Rightarrow$ by $\mathscr{D}$; obviously, $\mathscr{D}$ consists of those formulas of $\mathscr{C}$ which contain only $\Rightarrow$ and atoms. The following conditions have been given by [WAJ, 1935] (see also [CHU, 1956]); $\mathscr{D}$ is the smallest set such that:

(1) If $A \in \mathscr{D}$ and $A \Rightarrow B \in \mathscr{D}$, then $B \in \mathscr{D}$ (a rule of detachment already mentioned in 2.4.4);

(2) $A \Rightarrow (B \Rightarrow A) \in \mathscr{D}$;

(3) $A \Rightarrow (B \Rightarrow C) \Rightarrow ((A \Rightarrow B) \Rightarrow (A \Rightarrow C)) \in \mathscr{D}$;

(3) $(A \Rightarrow (B \Rightarrow C)) \Rightarrow ((A \Rightarrow B) \Rightarrow (A \Rightarrow C)) \in \mathscr{D}$;

(4) $((A \Rightarrow B) \Rightarrow A) \Rightarrow A \in \mathscr{D}$.

In [ŁUK, 1947] and [CHU, 1956, p. 111, Exercise 18.4], $\mathscr{D}$ is the smallest set satisfying (1), plus:

(5) $((A \Rightarrow B) \Rightarrow C) \Rightarrow ((C \Rightarrow A) \Rightarrow (D \Rightarrow A)) \in \mathscr{D}$.

2.4.6. *Tautologies in* $\neg, \bigwedge, \bigvee, \Rightarrow$

See the conditions of [JAS, 1948], quoted in [CHU, 1956, p. 167].

For every set of connections which contains at least $\Rightarrow$, hence in particular, for 2.4.4 through 2.4.6, there is a completeness lemma (proved by [HEN, 1949]).

2.5. Translation of the completeness lemma by means of weighted substitution

Let A be a connective formula and α a connection in A. Any weighted substitution replacing α by a formula representing α will yield a formula equideducible from A. However, the new formula need not represent the same connection as A. For example, let $A = +_2 a_3 a_2$, which represents the ternary connection $+_3$ always equal to $+$, and substitute for the binary connection $+_2$ the formula $B = +_2 a_2 a_2$, which also represents $+_2$; the substitution result is $+_2 a_2 a_2$, which represents the binary connection $+_2$ and not $+_3$.

Let us now return to the assertion of the completeness lemma in 2.4.1, which concerns the connections $\neg$ and $\bigvee$, and interpret it in terms of $\neg, \Rightarrow$, or $\neg, \bigwedge$, or other sets of connections.

2.5.1. *Tautologies in* $\neg, \Rightarrow$

Starting from a formula in $\neg$ and $\bigvee$, consider the weighted substitution replacing each $\bigvee$ by the formula $\Rightarrow \neg a_1 a_2$, which represents $\bigvee$. This amounts to ordinary substitution of the sequence $\Rightarrow \neg$ for each $\bigvee$. We get a formula in $\neg, \Rightarrow$ which is equideducible from the previous formula, in which each occurrence of $\Rightarrow$ is followed by an occurrence of $\neg$. Conversely, every formula in $\neg, \Rightarrow$ which satisfies the preceding condition is the substitution instance of a formula in $\neg$ and $\bigvee$.

It thus follows that the set of tautologies in $\neg, \Rightarrow$ *in which each* $\Rightarrow$ *is followed by* $\neg$ *is defined by translating the conditions of* 2.4.1 *as follows:*

(1′) for every A, $(\neg A) \Rightarrow (\neg A) \in \mathscr{C}$;
(2′) if $A \in \mathscr{C}$, then for every B we have $(\neg A) \Rightarrow B \in \mathscr{C}$;
(3′) if $A \in \mathscr{C}$, then $\neg\neg A \in \mathscr{C}$;
(4′) if $(\neg A) \Rightarrow B \in \mathscr{C}$, then $(\neg A) \Rightarrow (\neg\neg B) \in \mathscr{C}$;
(5′) if $(\neg A) \Rightarrow B \in \mathscr{C}$, then $(\neg B) \Rightarrow A \in \mathscr{C}$;
(6′) if $(\neg(\neg A \Rightarrow B)) \Rightarrow C \in \mathscr{C}$, then $(\neg(\neg B \Rightarrow A)) \Rightarrow C \in \mathscr{C}$;
(7′) if $(\neg(\neg A \Rightarrow B)) \Rightarrow C \in \mathscr{C}$, then $(\neg A) \Rightarrow (\neg B \Rightarrow C) \in \mathscr{C}$, and conversely;
(8′) if $(\neg(\neg(\neg A \Rightarrow B) \Rightarrow C)) \Rightarrow D \in \mathscr{C}$, then $(\neg(\neg A \Rightarrow (\neg B \Rightarrow C))) \Rightarrow D \in \mathscr{C}$ and conversely;

(9′) if $\neg A \in \mathscr{C}$ and $\neg B \in \mathscr{C}$, then $\neg(\neg A \Rightarrow B) \in \mathscr{C}$;

(10′) if $\neg A \Rightarrow \neg B \in \mathscr{C}$ and $\neg A \Rightarrow \neg C \in \mathscr{C}$, then $\neg A \Rightarrow \neg(\neg B \Rightarrow C) \in \mathscr{C}$.

To obtain the set $\mathscr{C}$ of *all* tautologies in $\neg$, $\Rightarrow$, it suffices to add conditions which will derive every tautology in $\neg$, $\Rightarrow$ from the preceding conditions, in which each $\Rightarrow$ is followed by a $\neg$. To the above ten conditions we add the following four:

(1″) if $A \Rightarrow A' \in \mathscr{C}$ and $A \in \mathscr{C}$, then $A' \in \mathscr{C}$ (rule of detachment for $\Rightarrow$);

(2″) $\neg\neg A \Rightarrow A \in \mathscr{C}$ and $A \Rightarrow \neg\neg A \in \mathscr{C}$;

(3″) if $A \Rightarrow A' \in C$, then $\neg A' \Rightarrow \neg A \in \mathscr{C}$;

(4″) if $A' \Rightarrow A \in \mathscr{C}$ and $B \Rightarrow B' \in \mathscr{C}$, then $(A \Rightarrow B) \Rightarrow (A' \Rightarrow B') \in \mathscr{C}$.

Indeed, let us start with a tautology in $\neg$, $\Rightarrow$. Each $\Rightarrow$ is followed by a subformula B of A. If B does not begin with $\neg$, substitute $\neg\neg B$ for B; rules (1′) through (10′) then show that $\mathscr{C}$ contains the substitution result A′ of A. Next, rules (2″) through (4″) show, step by step, that $A' \Rightarrow A \in \mathscr{C}$; rule (1″) finally gives $A \in \mathscr{C}$.

2.5.2. *Tautologies in* $\neg$, $\bigwedge$

Starting with a formula in $\neg$, $\bigvee$, we consider the weighted substitution of the formula $\neg \bigwedge \neg a_1 \neg a_2$ for each occurrence of $\bigvee$. This amounts to ordinary substitution of $\neg\neg\bigwedge$ for any $\bigvee$ preceded by another $\bigvee$, the first $\bigvee$ is replaced by $\neg\bigwedge$, and every occurrence of an atom is replaced by its negation. We get a formula in $\neg$, $\bigwedge$ which is equideducible from the preceding one, each $\bigwedge$ except the first is preceded by $\neg\neg$ (the first can only be preceded by $\neg$) and each atom is preceded by $\neg$. Conversely, every formula satisfying these conditions is a substitution instance of a formula in $\neg$, $\bigvee$. It therefore follows that *the set* $\mathscr{B}$ *of tautologies in* $\neg$, $\bigwedge$ *is defined by translating the conditions of* 2.4.1 (see Exercise 6) and adding the following four:

(1″) if $A \in \mathscr{B}$ and $\neg(A \bigwedge \neg A') \in \mathscr{B}$, then $A' \in \mathscr{B}$ (rule of detachment for $\neg$ and $\bigwedge$);

(2″) $\neg(\neg\neg A \bigwedge \neg A) \in \mathscr{B}$ and $\neg(A \bigwedge \neg\neg\neg A) \in \mathscr{B}$;

(3″) if $\neg(A \bigwedge \neg A') \in \mathscr{B}$, then $\neg(\neg A' \bigwedge \neg\neg A) \in \mathscr{B}$;

(4″) if $\neg(A \bigwedge \neg A') \in \mathscr{B}$ and $\neg(B \bigwedge \neg B') \in \mathscr{B}$, then $\neg((A \bigwedge B) \bigwedge \neg(A' \bigwedge B')) \in \mathscr{B}$.

2.5.3 *General case*

Let $\alpha_1, \ldots, \alpha_h$ be a finite number h of connections for which there exist a formula $N(a_1)$ in $\alpha_1, \ldots, \alpha_h$ representing the connective $\neg$ and a formula $V(a_1, a_2)$ representing $\bigvee$. Starting with a formula in $\neg$ and $\bigvee$, consider the weighted substitution replacing each occurrence of $\neg$ by the formula $N(a_1)$ and each occurrence of $\bigvee$ by the formula $V(a_1, a_2)$; we get an equideducible formula in $\alpha_1, \ldots, \alpha_h$.

Now denote the set of tautologies in $\alpha_1, \ldots, \alpha_h$ by $\mathcal{U}$. The *subset* of $\mathcal{U}$ consisting of the formulas formed by the indicated substitutions in the tautologies in $\neg$, $\bigvee$ *is defined by translating the conditions of* 2.4.1 *by means of* $N(a_1)$ *and* $V(a_1, a_2)$. Condition (1), for example, becomes: for every formula A, the formula $V(A, N(A)) \in \mathcal{U}$, and condition (2) becomes: if $A \in \mathcal{U}$ and B is any formula, then $V(A, B) \in \mathcal{U}$.

Let $\mathcal{U}'$ be the subset of $\mathcal{U}$ obtained by these translations of 2.4.1. All we need do to get all of $\mathcal{U}$ is to add conditions which will derive every tautology in $\alpha_1, \ldots, \alpha_h$ from those in $\mathcal{U}'$. To do this, associate with each tautology A in $\alpha_1, \ldots, \alpha_h$ the tautology A′ obtained by the following two consecutive weighted substitutions: first substitute for each α_i ($i = 1, \ldots, h$) a formula in $\neg$ and $\bigvee$ which represents α_i, and then substitute $N(a_1)$ for $\neg$ and $V(a_1, a_2)$ for $\bigvee$. Now let Eq (A, A′) denote the formula

$$V[N(V(N(A), N(A'))), N(V(A, A'))]$$

in $\alpha_1, \ldots, \alpha_h$, which is equal to $+$ when A and A′ take the same value. We now take the following conditions, which make it possible to derive every tautology A from a tautology $A' \in \mathcal{U}$:

(1″) if $\mathrm{Eq}(A, A') \in \mathcal{U}$ and $A \in \mathcal{U}$, then $A' \in \mathcal{U}$ (rule of detachment);

(2″) for each α_i ($i = 1, \ldots, h$) of arity k, if $\mathrm{Eq}(A_1, A'_1) \in \mathcal{U}, \ldots, \mathrm{Eq}(A_k, A'_k) \in \mathcal{U}$, then

$$\mathrm{Eq}(\alpha_i A_1 \ldots A_k, \alpha_i A'_1 \ldots A'_k) \in \mathcal{U};$$

(3″) for each $i = 1, \ldots, h$, let T_i be a formula in $\neg$, $\bigvee$ which represents α_i and U_i the formula obtained from T_i by substituting $N(a_1)$ for $\neg$ and $V(a_1, a_2)$ for $\bigvee$; then

$$\mathrm{Eq}(\alpha_i A_1, \ldots, A_k, U_i(A_1, \ldots, A_k)) \in \mathcal{U}.$$

2.6. Active, inactive index or atom; reduction, positive connection, alternating connection

2.6.1. An index $i \leqslant h$ will be called *inactive* in the h-ary connection α if the value $\alpha(x_1, \ldots, x_i, \ldots, x_h)$ does not depend on x_i: i.e.,

$$\alpha(x_1, \ldots, x_{i-1}, +, x_{i+1}, \ldots, x_h) = \alpha(x_1, \ldots, x_{i-1}, -, x_{i+1}, \ldots, x_h)$$

for all $x_1, \ldots, x_{i-1}, x_{i+1}, \ldots, x_h$.

If $\beta \vdash \alpha$ and i is inactive in α, there exists a connection γ whose inactive indices are i and the indices which are inactive in β, such that $\beta \vdash \gamma \vdash \alpha$. As an example, let γ be the disjunction of $\beta(x_1, \ldots, x_{i-1}, +, x_{i+1}, \ldots, x_h)$ and $\beta(x_1, \ldots, x_{i-1}, -, x_{i+1}, \ldots, x_h)$.

It thus follows that, given two connections α and β such that $\beta \vdash \alpha$, there exists a connection γ such that $\beta \vdash \gamma \vdash \alpha$, where each index inactive in α or in β is inactive in γ.

An atom a_i in a formula A is said to be *active* or *inactive* according as the index i is active or inactive in the connection represented by A.

2.6.2. Given a connection α, we define the *unary reduction of* α to be the unary connection whose value for $+$ $(-)$ is the same as that of α when all its atoms are assigned the value $+$ $(-)$. This is simply the connection represented by $\alpha a_1 \ldots a_1$ (all the atoms have been replaced by a_1). For example, the unary reduction of $\bigwedge$ is the connection ( ); the reduction of $\neg a \bigvee \neg b$ is negation $\neg$; the reduction of $\Rightarrow$ is the unary connection $+$; the reduction of $a \Rightarrow \neg b$ is $-$.

Let p be a positive integer; we define a *p-ary reduction of* α to be any p-ary connection represented by a formula built up from α followed by atoms chosen arbitrarily from among the a_i $(i = 1, 2, \ldots, p)$. In particular, if p is equal to the arity of α, every connection obtained by permutation of the atoms in $\alpha a_1 \ldots a_p$ is a p-ary reduction of α. For example, $b \Rightarrow a$ is a binary reduction of $a \Rightarrow b$.

If β is deducible from α, then the unary reduction of β is deducible from that of α. In general, for every p-ary reduction of β there exists a p-ary reduction of α from which the former is deducible.

2.6.3. A connection α is said to be *i-positive* (where i is an integer $\leqslant$ the arity) whenever substitution of $+$ for $-$ at rank i, arbitrary values being assigned to ranks $\neq i$, preserves the value of α or changes it from $-$ to $+$. The connection α is said to be *positive* if it is i-positive for every $i \leqslant$ the arity. In other words, the value taken by α remains the same or varies in the same sense as do the values of the system. The connections $+$, $-$,

$\bigwedge$ and $\bigvee$ are positive; negation $\neg$ is not; implication $\Rightarrow$ is 2-positive, but not 1-positive.

A connection is positive if and only if it is one of the connections $+$, $-$ or is representable by a formula in $\bigwedge$, $\bigvee$. Indeed, let α be positive, and consider every system of h values (where h = arity of α) which gives α the value $+$. For each of these systems, associate the atom a_i to each rank $i \leqslant h$ occupied by $+$. Take the conjunction of the a_i's for each system and then the disjunction of these conjunctions; the formula obtained will represent α. The two exceptions to this are the connections $-$ (since no system gives it the value $+$) and $+$ (for then we must also consider the system whose elements are all $-$, which does not correspond to any conjunction of atoms).

A formula all of whose connections are positive represents a positive connection. The dual of a positive connection is positive. The reductions of a positive connection are positive; the unary reduction will be ( ), except in the special cases of the connections $+$ and $-$.

2.6.4. A connection α is said to be *i-alternating* (where i is an integer $\leqslant$ the arity) if the values $\alpha(x_1, \ldots, x_{i-1}, +, x_{i+1}, \ldots)$ and $\alpha(x_1, \ldots, x_{i-1}, -, x_{i+1}, \ldots)$ are always distinct; in other words, if the value of x_i which gives α the value $+$ always exists and is unique, and is thus a function of the other values in the system. We see that α is then representable by the formula $a_i \Leftrightarrow (\alpha a_1 \ldots a_{i-1} + a_{i+1} \ldots)$.

A formula representing α is said to be *a_i-alternating*, where a_i is the atom of rank i under consideration. For example, the connection represented by $(a \bigwedge b \bigwedge c) \bigvee (\neg a \bigwedge \neg c) \bigwedge (\neg b \bigwedge \neg c)$ is c-alternating; by replacing c with $+$ we obtain $a \bigwedge b$, and the connection is representable by $c \Leftrightarrow (a \bigwedge b)$.

A connection α is said to be *alternating* whenever it is i-alternating for every active index i (see 2.6.1). In other words, change of value at a given rank either leaves the value of α unchanged or always changes that value. For example, the connections represented by a, $\neg a$, $a \Leftrightarrow b$, $a \Leftrightarrow \neg b$, and $(a \Leftrightarrow b) \Leftrightarrow c$ are alternating, and all their indices are active; $+(a)$ and $b \Leftrightarrow c$ are alternating and a is inactive.

A formula all of whose the connections are alternating represents an alternating connection. The alternating connections are exactly those which can be represented by a formula in $\Leftrightarrow$, and $-$. It is sufficient to substitute $+$ for the inactive atoms and then, for each active atom a, to replace the given formula $A(a)$ by $a \Leftrightarrow A(+)$ as already indicated, and then repeat the process (since $A(+)$ again represents an alternating connection). Any alternating connection of arity $\geqslant 1$ is representable by $\Leftrightarrow$ and $\neg$.

The dual of an alternating connection is alternating.

If a connection is alternating with an even number of active indices, then its unary reduction is $+(a)$ or $-(a)$. *Example:* $a \Leftrightarrow b$, $a \Leftrightarrow \neg b$. If the number of active indices is odd, the unary reduction is a or $\neg a$. *Example:* $(a \Leftrightarrow b) \Leftrightarrow \neg c$. Every reduction of an alternating connection is alternating. Every alternating connection with the reduction ( ) is representable by a formula in an even number of $\Leftrightarrow$; the same holds for the reduction $+(\ )$, with an odd number of $\Leftrightarrow$.

2.7. Set closed under formulation

Given a set of connections $\alpha, \beta, \ldots$, let us associate with this set all the connections represented by formulas in $\alpha, \beta, \ldots$, with arbitrary atoms; each such correspondence will be called a *formulation.* For example, starting from negation $\neg$ alone, we can obtain by formulation the binary connection represented by $\neg a_2$ and for each integer h the h-ary connection $\neg a_h$; starting with conjunction $\bigwedge$ alone, we obtain conjunctions $\bigwedge_h$ of arbitrary arities h, but we also get the ternary connection $a_2 \bigwedge a_3$, in which a_1 is inactive. From disjunction $\bigvee$ and the 0-ary connection $+$ we get the unary connection $+(a_1)$, which is represented by $a_1 \bigvee +$; while at the same time neither $\bigvee$ alone nor $+$ alone give $+(a_1)$.

A set $\mathscr{A}$ of connections is said to be *closed under formulation* if, whenever the connections $\alpha, \beta, \ldots$ belong to $\mathscr{A}$, then every connection represented by a formula in $\alpha, \beta, \ldots$ also belongs to $\mathscr{A}$. Every intersection of sets which are closed under formulation is also closed under formulation; the intersection of all sets closed under formulation which contain the connections $\alpha, \beta, \ldots$ will be called the set *generated* by $\alpha, \beta, \ldots$. This is the set of connections represented by formulas in $\alpha, \beta, \ldots$.

2.7.1. *A set $\mathscr{A}$ of connections is closed under formulation if and only if it satisfies the following two conditions:*

(1) *if* α *belongs to* $\mathscr{A}$, *then every reduction of* α (*see* 2.6.2) *belongs to* $\mathscr{A}$;

(2) *if* α *of arity i and* β *of arity j belong to* $\mathscr{A}$, *then the connection represented by* $\beta\alpha a_1 \ldots a_i a_{i+1} \ldots a_{i+j-1}$ *belongs to* $\mathscr{A}$.

It follows, for example, that the following sets are closed under formulation: the set of positive connections, the set of self-duals, the set of connections whose unary reduction is ( ) (represented by the atom a_1), and the set of connections which take the value + for the system all of whose elements are + (i.e., those whose unary reduction is ( ) or +( )).

2.7.2. *If a set is closed under formulation, then the set of dual connections is closed under formulation* (*see* 2.3.4).

Consider, for example, the set generated by the conjunction $\bigwedge$, which consists of conjunctions $\bigwedge_h$ of arbitrary arities h and of the connections obtained from some $\bigwedge_h$ by means of a bijective substitution f of indices, where the integers other than $f(1), \ldots, f(h)$ and less than $\max(f(1), \ldots, f(h))$ are inactive indices (for instance, 2 and 3 in $a_1 \bigwedge a_4$). Then the dual set is the set generated by the disjunction $\bigvee$ and built up similarly from $\bigvee_h$.

2.7.3. *Enumeration of the sets closed under formulation* [POS, 1941]

(1) The sets generated by the 0-ary connections (the values + and −) reduce to the empty set, the singleton {+}, the singleton {−}, and the pair {+, −}.

The sets generated by the 0-ary and unary connections are the following: The set $\mathscr{U}_1$ of all connections represented by an atom, which is generated by ( ) and, for each integer h, contains the h-ary connection a_h in which indices 1 to $h-1$ are inactive. The set $\mathscr{U}_2$ of tautologies, generated by +( ), and the dual set $\mathscr{U}_2^*$ of contradictions. The set $\mathscr{U}_3$ of atoms and their negations, generated by $\neg$. The union $\mathscr{U}_4$ of $\mathscr{U}_1$ and $\mathscr{U}_2$; its dual $\mathscr{U}_4^*$. The union $\mathscr{U}_5$ of $\mathscr{U}_2$ and its dual; the union $\mathscr{U}_6$ of $\mathscr{U}_1$, $\mathscr{U}_2$ and $\mathscr{U}_2^*$; the union $\mathscr{U}_7$ of $\mathscr{U}_1$, $\mathscr{U}_2$, $\mathscr{U}_2^*$ and $\mathscr{U}_3$ (these are all connections whose value is decided by the i-th value of the system, i being the only active index; the set is generated by $\neg$ and +( )). For each of the foregoing sets, and for those which follow, we shall distinguish between the variants obtained through including or excluding the 0-ary connectives + and −. For example, there exists a "pure" $\mathscr{U}_7$ generated by negation $\neg$ and the unary connection +( ) whose only value is +. We get a variant by adjoining the 0-ary connective $+$; the 0-ary connection $-$ is then represented by $\neg +$.

(2) The set $\mathscr{C}_1$ of all connections generated by negation $\neg$ and conjunction $\wedge$ (with or without the two 0-ary connections). The set $\mathscr{C}_2$ of connections whose unary reduction is $(\ )$ or $+(\ )$, generated by $\Rightarrow$ and $\wedge$; this set contains $a \vee \neg b$ but does not contain $\neg a$. The dual $\mathscr{C}_2^*$ with the unary reduction $(\ )$ or $-(\ )$. The intersection $\mathscr{C}_3$ of the two preceding sets, which contains all connections whose unary reduction is $(\ )$: this set is generated by $\wedge$, $\vee$ and $a \wedge (b \Rightarrow c)$; it contains $(a \Leftrightarrow b) \Leftrightarrow c$.

(3) The set $\mathscr{P}_1$ of positive connections, generated by $\wedge$, $\vee$, $+(\ )$, $-(\ )$. The set $\mathscr{P}_2$ of positive connections whose unary reduction is $(\ )$ or $+(\ )$; this set is generated by $\wedge$, $\vee$, $+(\ )$; its dual $\mathscr{P}_2^*$. The intersection $\mathscr{P}_3$ of the two preceding sets, generated by $\wedge$, $\vee$.

(4) The set $\mathscr{S}_1$ generated by $\vee$, $+(\ )$, $-(\ )$. The set $\mathscr{S}_2$ generated by $\vee$, $+(\ )$. The set $\mathscr{S}_3$ generated by $\vee$, $-(\ )$. The intersection $\mathscr{S}_4$ of the two preceding sets, generated by disjunction $\vee$ only. The four duals of the preceding sets, which are obtained by replacing $\vee$ with $\wedge$ and interchanging $+$ and $-$.

(5) The set $\mathscr{D}_1$ of self-dual connections. The set $\mathscr{D}_2$ of self-duals whose unary reduction is $(\ )$ (thus, the negation $\neg$ belongs to $\mathscr{D}_1$ but not to $\mathscr{D}_2$). The set $\mathscr{D}_3$ of positive self-duals; the connection $(a \Leftrightarrow b) \Leftrightarrow c$ belongs to $\mathscr{D}_2$ but not to $\mathscr{D}_3$. The unary reduction of a self-dual is necessarily $(\ )$; hence $\mathscr{D}_3$ is a subset of $\mathscr{D}_2$.

(6) The set $\mathscr{L}_1$ of alternating connections, generated by $\neg$ and $\Leftrightarrow$ (see 2.6.4). The set $\mathscr{L}_2$ of alternating connections whose unary reduction is $(\ )$ or $+(\ )$; this set is generated by $\Leftrightarrow$, and the connections $\neg$ and $a \Leftrightarrow \neg b$ belong to the first set but not to the second. The dual $\mathscr{L}_2^*$ (replace $+$ by $-$); this set is generated by $a \Leftrightarrow \neg b$ and does not contain $\Leftrightarrow$. The set $\mathscr{L}_3$ of alternating connections whose unary reduction is $(\ )$ or $\neg$; it does not contain $+(\ )$ but does contain $\neg$, and hence differs from the three preceding sets; $\mathscr{L}_3$ is the set of self-dual alternating connections, generated by $\neg$ and $(a \Leftrightarrow b) \Leftrightarrow c$. The intersection $\mathscr{L}_4$ of $\mathscr{L}_2$ and $\mathscr{L}_2^*$, the set of alternating connections whose unary reduction is $(\ )$; $\mathscr{L}_4$ is generated by $(a \Leftrightarrow b) \Leftrightarrow c$ and it differs from $\mathscr{D}_2$ and $\mathscr{D}_3$, since it does not contain $(a \wedge b) \vee (b \wedge c) \vee (c \wedge a)$.

(7) The other sets closed under formulation constitute eight infinite sequences, also obtained in [POS, 1941]; they are presented below in 2.7.9.

2.7.4. For every positive integer p, let $\mathscr{A}_p$ denote the set of connections α such that, whenever any p systems give α the value $-$, there exists i such that the i-th values of these p systems are $-$. It is clear that $\mathscr{A}_p$ contains $\mathscr{A}_{p+1}$ for each p.

In the case $p=1$, we get the connections which take the value $+$ for the system all of whose elements are $+$, i.e., the connections whose unary reduction is ( ) or $+$( ); hence $\mathscr{A}_1$ is the set $\mathscr{C}_2$ of the preceding section.

Let us denote by $\mathscr{A}_\omega$ the intersection of the $\mathscr{A}_p$, i.e., the set of connections α for which there exists an integer i such that every system which gives α the value $-$ has i-th value $-$. In other words, the *formulas which represent a connection in $\mathscr{A}_\omega$ are those which are deducible from one of their own atoms*, or those which are the disjunction of an atom and a formula; for example, $a \Rightarrow b$, $(a \wedge c) \vee b$.

The sets $\mathscr{A}_p$ (for each integer p) and $\mathscr{A}_\omega$ are closed under deduction. Indeed, if β is deducible from α, the systems which give the value $-$ to β also give the value $-$ to α.

The sets $\mathscr{A}_p$ (for each integer p) and $\mathscr{A}_\omega$ are closed under formulation. Indeed, they satisfy conditions (1) and (2) of 2.7.1. For condition (2), note that for the p systems which give the value $-$ to $\beta\alpha a_1 \ldots a_i a_{i+1} \ldots a_{i+j-1}$ there are two possibilities: (a) all of them give the value $-$ to $\alpha a_1 \ldots a_i$, so that they all have the value $-$ for a rank between 1 and i; (b) by replacing the first i values in each system by the resulting value for α, we can ensure that the p systems take the value $-$ for a rank between $i+1$ and $i+j-1$.

2.7.5. *The set $\mathscr{A}_\omega$ is generated by implication $\Rightarrow$ only.*

▷ Disjunction is generated by $\Rightarrow$, because $a \vee b$ is equideducible from $(a \Rightarrow b) \Rightarrow b$. Let a be an atom, B and C two formulas. By formulation, with the aid of $\Rightarrow$, we can go from the connections represented by $a \vee \mathrm{B}$ and $a \vee \mathrm{C}$ to the connection represented by $a \vee (\mathrm{B} \vee \mathrm{C})$. On the other hand, we need only substitute $a \vee \mathrm{B}$ for b in $b \Rightarrow a$ to obtain a formula representing $a \vee \neg \mathrm{B}$. As every connection of arity $\geqslant 1$ can be represented by a formula in $\neg$, $\vee$, the assertion is now proved. ◁

2.7.6. *All the $\mathscr{A}_p$ are distinct, and their intersection $\mathscr{A}_\omega$ is distinct from each of them.*

▷ The disjunction of the conjunctions $a_i \wedge a_j$ ($i \neq j$, $i, j = 1, 2, \ldots, p+1$)

belongs to $\mathscr{A}_p$ and not to $\mathscr{A}_{p+1}$. In fact, the systems of $p+1$ values which give this disjunction the value $-$ are the system all of whose elements are $-$ and the $p+1$ systems exactly one of whose elements is $+$. Hence, for any p systems giving the value $-$, there is a rank for which all of them have the value $-$. This is no longer true for the $p+1$ systems each having one value $+$. ◁

2.7.7. *Let α be an h-ary connection ($h \leqslant p$). Then α belongs to $\mathscr{A}_p$ if and only if α belongs to $\mathscr{A}_\omega$.*

▷ If α does not belong to $\mathscr{A}_\omega$, then for each rank $i=1, 2, \ldots, h$ there exists a system of h values which gives α the value $-$ and whose i-th value is $+$. We can then repeat one of these systems so as to bring their number up to p, and α will therefore not belong to $\mathscr{A}_p$. ◁

It follows that the connections in $\mathscr{A}_p - \mathscr{A}_\omega$ are at least $(p+1)$-ary; the example in 2.7.6 is of the smallest possible arity.

2.7.8. *If α belongs to $\mathscr{A}_p$, then every reduction of α of arity $\leqslant p$ belongs to $\mathscr{A}_\omega$* (reductions are defined in 2.6.2). This is a consequence of the fact that $\mathscr{A}_p$ is closed under formulation, and of the preceding assertion. On the other hand, $(a \wedge b) \vee \neg c$ has the unary reduction $+(\ )$, and binary reductions $a \vee \neg b$ and $\neg a \vee b$, all of which belong to $\mathscr{A}_\omega$. At the same time, the connection under consideration does not belong to $\mathscr{A}_2$, since the two systems $+-+$ and $-++$ each give it the value $-$ while there is no rank for which both have $-$.

If α is positive, then α belongs to $\mathscr{A}_p$ if and only if every reduction of arity $\leqslant p$ belongs to $\mathscr{A}_\omega$.

▷ By 2.7.8, it is sufficient to assume that α does not belong to $\mathscr{A}_p$ and to find a reduction of arity $\leqslant p$ which does not belong to $\mathscr{A}_\omega$. Take $h \leqslant p$ systems of values which give α the value $-$, such that for each rank the value $+$ appears at least once. Since α is positive, we replace some of the $+$ by $-$, at the same time requiring that in each rank there appears exactly one value $+$. Next, we reject all systems whose elements are all $-$ (if such occur). With each system we associate the set of ranks for which it takes the value $+$, and substitute for these the same atom; we thus get a reduction which is of arity $\leqslant p$ and does not belong to $\mathscr{A}_\omega$ (suggested by C. Benzaken). ◁

2.7.9. For each integer p and for ω, the following eight sets are closed under formulation: The set $\mathscr{A}_p$. The set $\mathscr{A}'_p$ of all positive connections in $\mathscr{A}_p$ (note that $\mathscr{A}'_1$ is the set $\mathscr{P}_2$ of 2.7.3). The set $\mathscr{A}''_p$ of all those connections in $\mathscr{A}_p$ whose unary reduction is ( ) (note that $\mathscr{A}''_1$ is the set $\mathscr{C}_3$ of 2.7.3). The intersection $\mathscr{A}'''_p$ of the two preceding sets (note that $\mathscr{A}'''_1$ is the set $\mathscr{P}_3$ of 2.7.3). Together these four sets, for each p and for ω, we have the four duals, which are also closed under formulation. Note that $\mathscr{A}'_\omega$ is generated by $+(\ )$ and $a\bigvee(b\bigwedge c)$: it is the set of positive connections deducible from one of their own atoms.

Post (1941) has proved that the *sets which are closed under formulation are precisely the sets* $\mathscr{U}$, $\mathscr{C}$, $\mathscr{P}$, $\mathscr{S}$, $\mathscr{D}$, $\mathscr{L}$ *of* 2.7.3, *along with the above eight sets* $\mathscr{A}$, *for each integer p and for* ω. It has also been proved that *each of these sets is generated by a finite number of connections.* From this it follows that given any infinite set $\mathscr{A}$ of connections, there is always a finite subset $\mathscr{B}$ of $\mathscr{A}$ such that every connection in $\mathscr{A}$ is represented by a formula in the connections of $\mathscr{B}$. Indeed, by the preceding result, the set $\mathscr{A}^*$, the closure of $\mathscr{A}$ under formulation, contains a finite subset which generates it, and this finite subset is itself generated by a finite number $\mathscr{B}$ of connections belonging to $\mathscr{A}$.

EXERCISES

1

(1) A connective formula in $\neg$, $\bigwedge$ and $\bigvee$ is said to be *unioccurrent* if each atom has a single occurrence. It is said to be *pseudopositive* if it can be obtained from a positive formula (see 2.6.3) by replacing certain atoms, at all their occurrences, with their negations. Show that every unioccurrent formula is equideducible from a pseudopositive formula; for example, $\neg((a \bigwedge \neg b) \bigvee c) \bigwedge d$ is equideducible from $(\neg a \bigvee b) \bigwedge \neg c \bigwedge d$.

Show that the converse is false; for example, the formula

$$(a \bigwedge b) \bigvee (b \bigwedge c) \bigvee (c \bigwedge a)$$

is positive but is not equideducible from any unioccurrent formula.

(2) Given a connective formula A and one of its atoms a, let $\underset{a}{\forall}\mathrm{A}$ denote any formula which is satisfied by those systems of values $+$ and $-$ assigned to the other atoms $b, c, \ldots$ which satisfy A for any value, $+$ or $-$, of a. Formulas $\exists\mathrm{A}$ are defined analogously, replacing "for any value" by "for at least one value". We have

$$\underset{a}{\forall}\mathrm{A}(b, c, \ldots) \dashv\vdash \mathrm{A}(+, b, c, \ldots) \bigwedge \mathrm{A}(-, b, c, \ldots)$$

and similarly with $\forall$ and $\exists$, $\bigwedge$ and $\bigvee$ interchanged. A connective formula A is said to be *permuting*, if finite sequences of operations $\forall$ and $\exists$ which differ only as to their order of application to A always yield equideducible formulas. For example, $a \bigwedge b$ is permuting, but $a \Leftrightarrow b$ is not, since $\underset{a}{\forall}\underset{b}{\exists}\mathrm{A}$ is $+$ while $\underset{b}{\exists}\underset{a}{\forall}\mathrm{A}$ is $-$. Show that every unioccurrent formula and, in general, every pseudopositive formula, is permuting.

Show that the formula $(a \bigwedge b) \bigvee (\neg a \bigwedge c)$ is permuting and is not equideducible from any pseudopositive formula.

Show that it is sufficient, in the definition of permuting formulas, to take only sequences of $\forall$ and $\exists$ involving all the atoms.

(3) Represent each sequence of h values $+$ and $-$ (h a positive integer) as a vertex of an h-dimensional cube, assigning, for example, to the i-th coordinate ($i = 1, \ldots, h$) the value 0 or 1 according as the i-th term of the sequence is $-$ or $+$. This sets up a bijective correspondence of each h-ary connection to its set $(+)$, i.e. the set of vertices on which the connection takes the value $+$, and to its set $(-)$, defined similarly (the complement of the preceding set). Show that a connective formula is pseudopositive if and only if there exists a vertex u belonging to the set $(+)$, such that, for every point x in $(+)$, all minimal paths joining u to x have their vertices in $(+)$ (a minimal path is formed from directed edges such that the end of one is the origin of the next and all edges have different directions). The same condition follows for the vertex opposite to u and the set $(-)$.

(4) Show that if the formula is permuting, then the sets $(+)$ and $(-)$ are *minimally connected*, i.e., for any two points u and v in the set $(+)$ there exists at least one minimal path joining u to v whose vertices are in $(+)$, and the same condition holds for the set $(-)$. Minimal connectedness may hold for $(+)$, but not for $(-)$; as an example, take $(a \bigwedge \neg b) \bigvee (b \bigwedge \neg c) \bigvee (c \bigwedge \neg a)$. In addition, there may also exist a path (not necessarily minimal) contained in $(+)$ between any two points in $(+)$, and the same for $(-)$, without these sets being minimally connected; an example is the conjunction of the preceding formula with the atom d. Finally, the following formula is not permuting, though its sets $(+)$ and $(-)$ are minimally connected:

$$(a \bigwedge b \bigwedge \neg d) \bigvee (\neg a \bigwedge \neg b \bigwedge d) \bigvee (b \bigwedge c \bigwedge d) \bigvee (\neg b \bigwedge \neg c \bigwedge \neg d).$$

(5) Show that an h-ary connection is permuting if and only if the sets $(+)$ and $(-)$ are minimally connected and neither of them reduces to $2h'$ vertices $(3 \leqslant h' \leqslant h)$ obtained consecutively from h' edges in different directions followed by h' edges in the opposite directions (the example at the end of (2) and parts (3) through (5) are due to [PAS, 1971]).

(6) Let A be a formula with atoms $a, b, \ldots$. Assume that there exists a disjunctive (see 2.3.3) and positive formula A_d with the atoms $a, b, \ldots$ and new atoms $a', b', \ldots$, in which no conjunction includes both a and a' or b and b', etc.; assume moreover that there exists a conjunctive and positive formula A_c, equideducible from A_d, in which no disjunction contains both a and a', or both b and b', etc. Finally, assume that when a' is replaced by $\neg a$, b' by $\neg b$, etc., one gets two formulas equideducible from A. Show that A is a permuting formula. For example, for $A = (a \wedge b) \vee (\neg a \wedge c)$, take

$$A_d = (a \wedge b) \vee (a' \wedge c) \vee (b \wedge c) \quad \text{and} \quad A_c = (a \vee c) \wedge (a' \vee b) \wedge (b \vee c).$$

Problem. If A is a permuting formula, we do not know whether A_d and A_c necessarily exist.

2

Expression of certain connections in terms of others (see [SHE, 1913]).

(1) Derive formulas equideducible from $\neg a$, $a \wedge b$ and $a \vee b$:

(a) using only the connection $|$ (joint denial), where $a \mid b = +$ if and only if $a = -$ or $b = -$;

(b) using only the connection $\|$,where $a \parallel b = +$ if and only if $a = b = -$.

(2) Show that $|$ and $\|$ are the only ones of the 16 binary connections from which one can obtain formulas equideducible from $\neg a$ and $a \wedge b$ (and thus formulas equideducible from any given formula).

Hint. First show that the binary connection α under consideration must yield a formula αaa equideducible from $\neg a$.

(3) Show that no formula in $\wedge$, $\vee$ is a tautology or a contradiction.

(4) Show that $\neg$ is not representable by a formula in $\wedge$, $\vee$, $\Rightarrow$.

(5) Show that $\wedge$ is not representable by a formula in $\vee$, $\Rightarrow$; verify that $\vee$ is representable by a formula in $\Rightarrow$, but not by a formula in $\Leftrightarrow$ (for $\vee$ is not alternating). Neither is $\Rightarrow$ representable by $\Leftrightarrow$, nor $\Leftrightarrow$ by $\Rightarrow$.

3

Even or odd occurrence (see [HER, 1930]). Let A be a connective formula in $\neg$, $\wedge$, $\vee$, and t an occurrence of an atom a. We call an occurrence *even* if, in the (total) ordering of the terms which dominate t, there is an even number (possibly zero) of occurrences of negation $\neg$; an occurrence is *odd* when there is an odd number of $\neg$'s.

(1) For the occurrence t of a, substitute first the atom b, then the formula $a \wedge b$, and then $a \vee b$, thus obtaining from $A_a = A$ formulas A_b, $A_{a \wedge b}$, and $A_{a \vee b}$. Show that if the occurrence is even:

$$A_{a \wedge b} \text{ is equideducible from } A_a \wedge A_b,$$
$$A_{a \vee b} \text{ is equideducible from } A_a \vee A_b.$$

What happens if the occurrence is odd?

Example. $(\neg a \vee b) \wedge a$, and t is the first occurrence of a.

In the above notation, show that, for every occurrence of a, the following deductions hold: $A_a \wedge A_b \vdash A_{a \wedge b}$ and $A_{a \vee b} \vdash A_a \vee A_b$ (replace a and b by $+$ and $-$ in all possible ways). The same holds when several, even all occurrences of a are considered.

Consequently, the above equideductions are valid when all occurrences of a are even and the substitutions are simultaneous.

(2) Let $A = A_a$ be a formula in $\neg, \bigwedge, \bigvee$ which contains an atom a and in which, furthermore, certain atoms may be replaced by the values $+$ or $-$ (in such a manner that a tautology, for example, can contain neither even nor odd occurrences). Let us assume that, for some occurrence of a, and for an atom b which does not appear in A, we have the equideduction $A_{a\wedge b} \dashv\vdash A_a \bigwedge A_b$. Show that then $A_{a\vee b} \dashv\vdash A_a \bigvee A_b$. More precisely, there exists a formula equideducible from A_b in which b can have no odd occurrence. Distinguish between the cases in which the occurrence of a being considered is even or odd; in the latter case, A_b is equideducible from A_a and from the formulas obtained when b is replaced by $+$ or $-$.

(3) If the equideduction $A_{a\wedge b} \dashv\vdash A_a \bigwedge A_b$ holds for every occurrence of a (and b not appearing in A), then A is equideducible from a formula in which all occurrences of a (if there are any) are even. Note that the equideduction under consideration is not preserved if $+$ or $-$ is substituted for another occurrence of a. Take, for example, $\neg a \bigvee a \bigvee \neg a$, which satisfies the equideduction for the first occurrence of a, while the substitution instance $\neg a \bigvee a \bigvee \neg +$ does not.

(4) Do the same for simultaneous substitution for all occurrences of a.

4

(1) Starting from the primitive conditions of [RUS, 1908] or [RUS-WHI, 1910], i.e., conditions (1) through (5) of 2.4.2, plus

$$\neg(A \bigvee (B \bigvee C)) \bigvee (B \bigvee (A \bigvee C)) \in \mathscr{A},$$

derive conditions (1) through (10) of the completeness lemma 2.4.1.

(2) Deduce the condition $\neg(A \bigvee (B \bigvee C)) \bigvee (B \bigvee (A \bigvee C)) \in \mathscr{A}$ from conditions (1) through (5) of 2.4.2. Show first that

$$\neg(A \bigvee (B \bigvee C)) \bigvee ((B \bigvee (A \bigvee C)) \bigvee A) \in \mathscr{A}$$

and

$$\neg((B \bigvee (A \bigvee C)) \bigvee A) \bigvee (B \bigvee (A \bigvee C)) \in \mathscr{A}.$$

5

Following the scheme of 2.4.1, show that the tautologies in $\neg, \bigwedge$ constitute the smallest set $\mathscr{B}$ such that:

(1) for every formula A, $\neg(A \bigwedge \neg A) \in \mathscr{B}$;
(2) if $\neg A \in \mathscr{B}$, then $\neg(A \bigwedge B) \in \mathscr{B}$;
(3) if $A \in \mathscr{B}$, then $\neg\neg A \in \mathscr{B}$;
(4) if $A \in \mathscr{B}$ and $B \in \mathscr{B}$, then $A \bigwedge B \in \mathscr{B}$;
(5) if $\neg(A \bigwedge B) \in \mathscr{B}$, then $\neg(B \bigwedge A) \in \mathscr{B}$;
(6) if $\neg((A \bigwedge B) \bigwedge C) \in \mathscr{B}$, then $\neg((B \bigwedge A) \bigwedge C) \in \mathscr{B}$;
(7) if $\neg((A \bigwedge B) \bigwedge C) \in \mathscr{B}$, then $\neg(A \bigwedge (B \bigwedge C)) \in \mathscr{B}$ and conversely;
(8) if $\neg(((A \bigwedge B) \bigwedge C) \bigwedge D) \in \mathscr{B}$, then $\neg((A \bigwedge (B \bigwedge C)) \bigwedge D) \in \mathscr{B}$ and conversely;
(9) if $\neg(A \bigwedge B) \in \mathscr{B}$, then $\neg(A \bigwedge \neg\neg B) \in \mathscr{B}$;
(10) if $\neg(A \bigwedge \neg B) \in \mathscr{B}$ and $\neg(A \bigwedge \neg C) \in \mathscr{B}$, then $\neg(A \bigwedge \neg(B \bigwedge C)) \in \mathscr{B}$.

6

Starting from a formula in $\neg, \bigvee$ which contains at least one $\bigvee$, we get an equideducible formula in $\neg$, $\bigwedge$ by replacing each occurrence of an atom by its negation and each $\bigvee$ by $\neg\bigwedge$ or $\neg\neg\bigwedge$, depending on whether it is not or is preceded by another $\bigvee$ (see 2.5.2).

(1) Show that a formula in $\neg$, $\bigwedge$ can be obtained in this way from a formula in $\neg$, $\bigvee$ if and only if (a) it contains a $\bigwedge$ and (b) each occurrence of an atom is immediately preceded by $\neg$ and each $\bigwedge$ is preceded by $\neg\neg$ (or $\neg$ for the first occurrence of $\bigwedge$).

(2) Show that the set of tautologies in $\neg$, $\bigwedge$ is the smallest set $\mathscr{B}$ such that:

(i) if $A \in \mathscr{A}$ (the smallest set closed under conditions (1) through (10) of 2.4.1), then the transform of A belongs to $\mathscr{B}$;

(ii) $\mathscr{B}$ satisfies the four conditions (1″) to (4″) of 2.5.2.

7

Let $\mathscr{C}$ be a set of formulas in $\neg$, $\Rightarrow$ such that:

(1) for every formula A, $A \Rightarrow A \in \mathscr{C}$;
(2) if $A \in \mathscr{C}$, then $B \Rightarrow A \in \mathscr{C}$;
(3) if $A \in \mathscr{C}$, then $\neg\neg A \in \mathscr{C}$;
(4) if $A \in \mathscr{C}$ and $\neg B \in \mathscr{C}$, then $\neg(A \Rightarrow B) \in \mathscr{C}$;
(5) if $A \Rightarrow B \in \mathscr{C}$, then $\neg\neg A \Rightarrow B \in \mathscr{C}$;
(6) if $\neg A \Rightarrow C \in \mathscr{C}$ and $B \Rightarrow C \in \mathscr{C}$, then $(A \Rightarrow B) \Rightarrow C \in \mathscr{C}$;
(7) if $A \Rightarrow (\neg B \Rightarrow C) \in \mathscr{C}$, then $\neg(A \Rightarrow B) \Rightarrow C \in \mathscr{C}$;
(8) if $A \Rightarrow B \in \mathscr{C}$, then $A \Rightarrow \neg\neg B \in \mathscr{C}$;
(9) if $A \Rightarrow B \in \mathscr{C}$ and $A \Rightarrow \neg C \in \mathscr{C}$, then $A \Rightarrow \neg(B \Rightarrow C) \in \mathscr{C}$;
(10) if $A \Rightarrow (B \Rightarrow C) \in \mathscr{C}$, then $B \Rightarrow (A \Rightarrow C) \in \mathscr{C}$;
(11) if $A \Rightarrow (B \Rightarrow C) \in \mathscr{C}$, then $\neg(A \Rightarrow \neg B) \Rightarrow C \in \mathscr{C}$ and conversely;
(12) if $A \Rightarrow (B \Rightarrow (C \Rightarrow D)) \in \mathscr{C}$, then $A \Rightarrow (\neg(B \Rightarrow \neg C) \Rightarrow D) \in \mathscr{C}$ and conversely;
(13) if $A \Rightarrow (B \Rightarrow (C \Rightarrow D)) \in \mathscr{C}$, then $A \Rightarrow (C \Rightarrow (B \Rightarrow D)) \in \mathscr{C}$.

(1) By the method of 2.4.1, deduce from conditions (10) through (13) that any formula A in $\neg$, $\Rightarrow$ that does not begin with $\neg$ can be expressed as

$$A_1 \Rightarrow (A_2 \Rightarrow (A_3 \Rightarrow (\cdots \Rightarrow (A_h \Rightarrow B) \cdots)))),$$

where B is either an atom or begins with $\neg$; moreover, A belongs to $\mathscr{C}$ if and only if the formula

$$A_{i_1} \Rightarrow (A_{i_2} \Rightarrow (A_{i_3} \Rightarrow (\cdots \Rightarrow (A_{i_h} \Rightarrow B) \cdots))))$$

belongs to $\mathscr{C}$ for every permutation $(i_1, \ldots, i_h)$ of $1, \ldots, h$.

Verify that condition (13) follows from the other three.

(2) Show that the set of tautologies in $\neg$, $\Rightarrow$ is the smallest set $\mathscr{C}$ satisfying conditions (1) through (12).

8

(1) From conditions (1), (2) and (3) of 2.4.4, deduce condition (10) of the same section i.e.,

$$(A \Rightarrow B) \Rightarrow ((B \Rightarrow C) \Rightarrow (A \Rightarrow C)) \in \mathscr{C}$$

(use the following two lemmas: if $A \Rightarrow B \in \mathscr{C}$ and $B \Rightarrow C \in \mathscr{C}$, then $A \Rightarrow C \in \mathscr{C}$; moreover, $(B \Rightarrow C) \Rightarrow ((A \Rightarrow B) \Rightarrow (A \Rightarrow C)) \in \mathscr{C}$).

(2) From the same conditions, deduce:

$$(A \Rightarrow (B \Rightarrow C)) \Rightarrow (B \Rightarrow (A \Rightarrow C)) \in \mathscr{C}.$$

(This condition is used in [FRE, 1879].)

NOTE

* The more usual terms in English are "connective" and "propositional formula". However, the author expressly uses French "connection" and "formule connective" rather than "connectif" and "formule propositionnelle", and moreover "connection" has also been used in this meaning in English (see, e.g., R. Carnap, *Formalization of Logic*, Harvard University Press, 1947; H. B. Curry, *Foundations of Mathematical Logic*, McGraw-Hill, 1963). – Trans.

CHAPTER 3

RELATION, MULTIRELATION; OPERATOR AND PREDICATE

From this chapter on we shall employ all the usual set-theoretic axioms and definitions, including the axiom of infinity, i.e., the existence of a set containing 0 which, if it contains a, also contains $a+1=a\cup\{a\}$. Explicit mention of the axiom of choice and of the ultrafilter axiom (every filter is contained in a finer ultrafilter) will be made whenever these are used.

3.1. RELATION, MULTIRELATION, BASE, ARITY; CONCATENATION; RESTRICTION AND EXTENSION

Let E be a set and m a natural number. A sequence of m elements $x_1, \ldots, x_m$ is called an m-tuple over E; for $m=2$ we shall speak of a *pair*, and for $m=3$, a *triple*. An *m-ary relation* on E is a function R which associates with every m-tuple a value $R(x_1, \ldots, x_m) = +$ or $-$. The set E is the *base* of R; the integer m is its *arity*. For $m=1, 2, 3$ we shall speak of a *unary, binary, ternary* relation. For $m=0$ we adopt the convention that there exist two 0-ary relations on E, denoted by (E, +) and (E, −); the *value* of the 0-ary relation is + or −, respectively. The base may be empty, $E=\emptyset$; for this case we adopt the convention that there exists exactly one m-ary relation for each positive m, plus two 0-ary relations, $(\emptyset, +)$ and $(\emptyset, -)$.

Examples. For every element a of E, the *singleton* of a is the unary relation over base E, equal to + for a and − for any other element.

The ordering of the natural numbers is the binary relation I defined for natural numbers x_1, x_2 as $I(x_1, x_2) = +$ for $x_1 \leqslant x_2$ and $I(x_1, x_2) = -$ for $x_1 > x_2$.

A group is a ternary relation $R(x_1, x_2, x_3)$ which is equal to + if and only if $x_1 . x_2 = x_3$, where the dot denotes the composition law of the group.

When E is a finite set containing exactly p elements, one easily sees that there exist 2^{p^m} m-ary relations over E (with the convention that $0^0=1$).

Instead of the notation $x_1, x_2, x_3, \ldots$, we shall sometimes find it more convenient to use the notation $x, y, z, \ldots$.

A finite sequence M of relations $R_1, \ldots, R_h$ with common base E, indexed by the integers $1, \ldots, h$, will be called a *multirelation* over E. The *arity* of M is the sequence $\mu = (m_1, \ldots, m_h)$ of the arities of $R_1, \ldots, R_h$; we shall call M a *μ-ary* multirelation. The set E is the *base* of M. The sequence of relations may be empty, in which case the multirelation reduces to the base E. For $h = 2, 3, \ldots$, we speak of *birelations, trirelations*, etc.

A relation or multirelation is said to be *finite, infinite, denumerable* or *of cardinality* α according to whether its base is finite, infinite, denumerable or of cardinality α.

Examples. The ordering R_1 of the natural numbers and the relation $R_2(x_1, x_2) = +$ when $x_2 = x_1 + 1$ form a multirelation (R_1, R_2) on the integers.

An ordered group is a multirelation (R_1, R_2) formed by a ternary group relation R_1 and a binary ordering relation R_2 compatible with the group R_1.

We shall use the notation R, S, T,... instead of R_1, R_2, R_3... when it proves more convenient.

Given two multirelations M and N with common base E, the *concatenation* of M and N, denoted by MN, is defined as the sequence made up of the relations in M followed by those in N (with the indices of the latter increased by the number of terms in M).

3.1.1. Let F be a subset of E; the *restriction* of R to F, denoted by $R \mid F$, is defined as the relation on F which takes the same values as R for every m-tuple over F. The restriction of the 0-ary relation (E, +) is (F, +); that of (E, $-$) is (F, $-$). If $F = \emptyset$, we get the unique m-ary relation with the empty base for positive m, and the relation $(\emptyset, +)$ or $(\emptyset, -)$ for $m = 0$.

Given a relation R over E and a superset $E^* \supseteq E$, an *extension* of R to E^* is any relation R^* with base E^* whose restriction to E is R.

Let R *and* R' *be two m-ary relations having a common base* E. *If* $R \mid F = R' \mid F$ *for every subset* F *of* E *containing at most m elements, then* $R = R'$.

Indeed, let $x_1, \ldots, x_m$ be an m-tuple over E; the set $F = \{x_1, \ldots, x_m\}$ contains at most m elements (some of the x_i may be equal), so from $R \mid F = R' \mid F$ it follows that

$$R(x_1, \ldots, x_m) = R'(x_1, \ldots, x_m).$$

We define the *restriction* of a multirelation $M=(R_1,\ldots,R_h)$, denoted by $M \mid F$, as the multirelation $(R_1 \mid F,\ldots,R_h \mid F)$. An *extension* of M to $E^* \supseteq E$ is any multirelation M^* with base E^* whose restriction to E is M, or, equivalently, any sequence $(R_1^*,\ldots,R_h^*)$ in which each R_i^* is an extension of R_i to E^* (for $i=1,\ldots,h$).

Let M *be of arity* $(m_1,\ldots,m_h)$ *and let m be the maximum of* $m_1,\ldots,m_h$*; if* $M \mid F=M' \mid F$, *for every subset* F *of* E *containing at most m elements, then* $M=M'$.

3.1.2. (1) *Consider a set* $\mathscr{M}$ *of multirelations which are pairwise compatible, in the sense that if* M *and* M′ *belong to* $\mathscr{M}$ *and their bases are* E *and* E′, *the restrictions* $M \mid E\cap E'$ *and* $M' \mid E\cap E'$ *are identical. Then there exists a multirelation on the union of the bases which is an extension of each multirelation in* $\mathscr{M}$.

(2) *In particular, if the set* $\mathscr{E}$ *of the bases of the multirelations in* $\mathscr{M}$ *is a directed set, i.e., if for* $E, E'\in\mathscr{E}$ *there exists* $E''\in\mathscr{E}$ *such that* $E''\supseteq E\cup E'$, *and if* $M' \mid E=M$ *whenever the base* E *of* M *is a subset of the base* E′ *of* M′, *then there exists a common extension of the multirelations in* $\mathscr{M}$.

(3) *Returning to the general case of* (1), *if every set of m elements, where m is the maximum arity of the multirelations, is contained in the base of a multirelation in* $\mathscr{M}$, *then the above extension is unique.*

▷ (1) Assume that the M in $\mathscr{M}$ are all m-ary relations, and let $x_1,\ldots,x_m$ be an m-tuple over the union of their bases. Now, either there exists a base E containing all the x_i $(1\leqslant i\leqslant m)$, or no base of a multirelation in $\mathscr{M}$ contains all the x_i. In the first case, suppose $M\in\mathscr{M}$ has base E. Consider the value $M(x_1,\ldots,x_m)$; by hypothesis, this value does not depend on the choice of M. In the second case, associate with the m-tuple, say, the value $+$.

Part (2) is a special case of (1). For given M with base E and M′ with base E′, both belonging to $\mathscr{M}$, there exists M″ in $\mathscr{M}$ with base $E''\supseteq E\cup E'$ such that $M'' \mid E\cap E'=M \mid E\cap E'=M' \mid E\cap E'$.

Part (3) follows from 3.1.1. ◁

3.1.3. *Coherence Lemma*

Let $\mathscr{E}$ *be a set of sets* E, *over each of which is given a nonempty finite set* U_E *of multirelations with base* E. *Suppose that the following two conditions hold:*

(1) *$\mathscr{E}$ is a directed set, i.e., if* E *and* $E' \in \mathscr{E}$, *then there exists* $E'' \in \mathscr{E}$ *such that* $E'' \supseteq E \cup E'$.

(2) *If* E *and* $E' \in \mathscr{E}$ *and* $E \subseteq E'$, *then the restriction to* E *of every multirelation in* $U_{E'}$ *gives a multirelation in* U_E.

Then there exists a multirelation N *over the union of the sets* E *in* $\mathscr{E}$ *such that for every* E *the restriction* $N \mid E$ *belongs to* U_E.

Proof, using the ultrafilter axiom. ▷ Denote the union of the sets E by D, and with each U_E associate the set V_E of extensions to D of the multirelations in U_E. The supersets of the sets V_E constitute a filter over the set of multirelations of the given arity with base D. Indeed, no V_E is ever empty, since by hypothesis no U_E is ever empty. Given two sets E and E′ in $\mathscr{E}$, there exists E″ in $\mathscr{E}$ such that $E'' \supseteq E \cup E'$ and the intersection $V_E \cap V_{E'}$ is a superset of $V_{E''}$, and therefore an element of the filter. Now consider a finer ultrafilter. For each E in $\mathscr{E}$, partition the multirelations M in V_E into a finite number of equivalence classes, defining M to be equivalent to M′ if $M \mid E = M' \mid E$. Exactly one of these classes is an element of the ultrafilter. Let N_E be the common restriction of the multirelations in this class to E. The fact that V_E is an element of the ultrafilter implies that there exist multirelations in V_E whose restriction to E is N_E, and hence $N_E \in U_E$. It remains to be shown that there exists a common extension N of the N_E whose base is D. Now, if E, $E' \in \mathscr{E}$ and $E \subseteq E'$, it follows from the construction of the ultrafilter that there exists a multirelation with base D having both the restriction N_E over E and the restriction $N_{E'}$ over E′. Therefore N_E is a restriction of $N_{E'}$; the extension N of the N_E exists by 3.1.2 (2). ◁

3.1.4. Conversely, *the Coherence Lemma implies* (*and is therefore equivalent to*) *the ultrafilter axiom*

▷ Let E be a set, P(E) the power set of E and $\mathscr{F}$ a filter over E. With any finite subset F of P(E) which is closed under $\cap$, $\cup$ and complementation with respect to E, we associate the nonempty set U_F of unary relations X with base F which satisfy the following conditions:

for every $A \in F$ and $A \in \mathscr{F}$, we have $X(A) = +$;
for every $A \in F$ such that $E - A \in \mathscr{F}$, we have $X(A) = -$;
for every $A \in F$ and $B \in F$ such that $B = E - A$, we have $X(A) \neq X(B)$;
for $A \in F$, $B \in F$ and $A \cap B \in F$, if $X(A) = X(B) = +$, then $X(A \cap B) = +$;
for $A \in F$, $A' \in F$ and $A \subseteq A'$, if $X(A) = +$, then $X(A') = +$.

These conditions can be satisfied by taking an element u in the intersection of the subsets of E which belong both to F and to $\mathscr{F}$, and setting $X(A) = +$ or $-$ according to whether $u \in A$ or not. Using these conditions, we apply the Coherence Lemma to the sets U_F. It follows that there exists a unary relation over P(E) whose restriction to each F belongs to U_F. The subsets of E that give the value + to this relation constitute an ultrafilter over E which is finer than $\mathscr{F}$. ◁

3.1.5. Another version of the Coherence Lemma is due to [RAD, 1949]. Consider disjoint finite sets U and, for every finite set I of sets U, let f_I be a choice function which associates with each U in I an element $f_I(U)$ of U. Then there exists a choice function f over all the sets U, such that, for any finite set I of U's, there exists $J \supseteq I$ with f_J a restriction of f. This assertion, combined with the axiom of choice for finite sets, is equivalent to the Coherence Lemma (see [BEN, 1970]).

3.2. Isomorphism, automorphism, embedding, age

Let f be a bijective mapping of E onto a set E′; the relation R′ on E′, defined by

$$R'(f(x_1), \ldots, f(x_m)) = R(x_1, \ldots, x_m)$$

for any $x_1, \ldots, x_m \in E$ is said to be *isomorphic* to R, or, more precisely, the *image* of R under the *isomorphism* f; we write $R' = f(R)$.

If R is 0-ary, then either $R = (E, +)$ and its image is $f(R) = (E', +)$, or $R = (E, -)$ and its image is $(E', -)$.

An *automorphism* of R is an isomorphism of R onto R.

The *empty mapping*, denoted by $f_\emptyset$, is the mapping satisfying the following conventions:

$f_\emptyset$ maps the empty set onto itself, and is bijective; $f_\emptyset$ is the restriction to the empty set of any mapping, in other words, every mapping is an extension of $f_\emptyset$.

$f_\emptyset$ preserves every relation with empty base.

Isomorphism and automorphism extend to multirelations: f transforms

$$M = (R_1, \ldots, R_h) \quad \text{into} \quad f(M) = (f(R_1), \ldots, f(R_h)).$$

3.2.1. *Let* E *be a set,* f *a permutation of* E, *and* F *a finite subset of* E. *There exists a sequence of distinct elements* a_i $(i=1,\ldots,h)$ *of* F *such that, for any element* x *of* F, *the image* $f(x)$ *is identical to the image of* x *under the product of the transpositions* $(a_i, f(a_i))$ *from* $i=1$ *to* h.

▷ Partition F into maximal subsets (with respect to inclusion) of the form $u_1, u_2, \ldots, u_k$ (k a positive integer) where $u_2=f(u_1), \ldots, u_k=f(u_{k-1})$. With each of these maximal subsets of F we associate, in the following order, the sequence of transpositions $(u_k, f(u_k))$, $(u_{k-1}, u_k), \ldots, (u_1, u_2)$ if $f(u_k)\neq u_1$, or the sequence $(u_{k-1}, u_k), \ldots, (u_1, u_2)$ if $f(u_k)=u_1$, that is to say, if the maximal subset in question extends to a complete cycle of the permutation f. We see that each element x of the subset in question is mapped onto $f(x)$; with this subset, therefore, we associate the sequence of distinct elements $u_k, u_{k-1}, \ldots, u_1$. With the whole set F we associate the sequence formed from all the above partial sequences, in any order. ◁

3.2.2. *Let* R *be a relation with base* E *and* f *a permutation of* E. *Suppose that for every element* a *of* E *the transposition* $(a, f(a))$ *is an automorphism of* R. *Then* f *is an automorphism of* R (private communication from G. Lopez).

▷ Let n be the arity of R; if f changes R, there exists a sequence of n elements a_i of E $(i=1,\ldots,n)$ such that $R(a_1,\ldots,a_n)\neq R(f(a_1),\ldots,f(a_n))$. By the previous assertion (3.2.1), there exist an integer h and a sequence $b_1,\ldots,b_h$ of elements $a_i (i=1,\ldots,n)$ such that the product of the transpositions $(b_j, f(b_j))$ from $j=1$ to h maps each a_i onto $f(a_i)$, and so changes R. At least one of these transpositions changes R. ◁

3.2.3. We shall say that a multirelation R is *embeddable* in S, writing $R \leqslant S$, if there exists a restriction of S which is isomorphic to R. Embeddability is a pre-ordering (reflexive and transitive relation) over multirelations of fixed arity. It is compatible with isomorphism: if R′ is isomorphic to R, S′ is isomorphic to S, and $R \leqslant S$, then $R' \leqslant S'$.

If R and S are multirelations with finite bases, the double embedding $R \leqslant S \leqslant R$ is equivalent to isomorphism of R and S. This is false for infinite bases; for example, any chain R with finite or denumerable base is embeddable in the chain Q of rational numbers: simply enumerate the elements of the base of R, and note that, if the first n elements of R have been mapped into Q, the $(n+1)$-th element can also be mapped onto a suitable rational number, preserving all order relations, since between any two rationals there exists a third, and there is a rational number before and after any rational number. It follows that the chain Q+1 derived from Q by adjoining a "last" element following all the rationals satisfies $Q \leqslant Q+1 \leqslant Q$, while Q+1 is not isomorphic to Q. Another example: let N denote the chain of natural numbers, N^- the reverse chain, isomorphic to the chain of negative integers; let $N(N^-)$ denote the chain obtained by replacing every natural number in N by a chain isomorphic to N^-. One sees that

$$N(N^-) \leqslant 1 + N(N^-) \leqslant N(N^-),$$

while these orderings are certainly not isomorphic.

3.2.4. Let R be a multirelation; the set $\mathscr{R}$ of all finite restrictions of R (considered only up to isomorphism) is called the *finite-age* of R; it satisfies the following conditions:

(1) if $A \in \mathscr{R}$ and $B \leqslant A$, then $B \in \mathscr{R}$;

(2) if $A, B \in \mathscr{R}$, there exists $C \in \mathscr{R}$ such that $C \geqslant A$ and $C \geqslant B$.

Note that the restrictions of R to $\leqslant p$ elements (p a natural number) satisfy (1), but not necessarily (2); the restrictions to $\geqslant p$ elements satisfy (2) but not (1).

Given a set $\mathscr{R}$ of finite multirelations satisfying (1) and (2), there exists a finite or denumerable multirelation whose finite restrictions form (a set isomorphic to) the set $\mathscr{R}$.

Given a multirelation R, the class of multirelations with the same (up to isomorphism) finite restrictions as R is called the *age* of R. For example, the infinite chains form an age; the corresponding finite restrictions are the finite chains.

A set of finite multirelations closed under embeddability is a finite-age if and only if it is not the union of two different sets closed under embeddability.

Assuming the axiom of choice: *For any set $\mathscr{A}$ of finite multirelations closed under embedding, and any finite-age $\mathscr{R}$ contained in $\mathscr{A}$, there exists a finite-age which is maximal among those containing $\mathscr{R}$ and contained in $\mathscr{A}$.*

For a generalization of the concept of age, see [CUS-PAB, 1970].

3.2.5. If we admit the ultrafilter axiom, then:

Given a multirelation M *with infinite base* E *and a superset* E* *of* E, *there exists an extension* M* *of* M *to* E* *such that every finite restriction of* M* *is isomorphic to a restriction of* M.

More precisely, *for every finite subset* F *of* E* *there exists an isomorphism of* M* | F *onto a restriction of* M *leaving fixed every element of* F∩E.

▷ For each finite subset F of E* let U_F be the set of extensions X of M to E* for which there exists an isomorphism of X | F onto a restriction of M leaving fixed every element of F∩E. To obtain an extension X, first take a bijective mapping f of F onto any subset of E such that $f(x)=x$ for every $x \in F \cap E$. Next, consider the inverse image $f^{-1}(M \mid f(F))$, whose restriction to F∩E is equal to that of M. Finally, take an extension over E* which is common to M and to the preceding (inverse) image. It follows that for all F the set U_F is nonempty. Moreover, for $F' \supseteq F$ we have $U_{F'} \subseteq U_F$, and so for all finite subsets F and G of E* we have $U_{F \cup G} \subseteq$ $\subseteq U_F \cap U_G$. Therefore, the sets U_F and their supersets constitute a filter over the multirelations which are extensions of M to E*. Take a finer ultrafilter. Over each F there are only finitely many multirelations having the same arity as M; for exactly one of these, say N_F, the set of extensions common to M and to N_F is an element of the ultrafilter. For each $F' \supseteq F$ the multirelation $N_{F'}$ is an extension of N_F. Hence there exists exactly one multirelation over E*, which is a common extension of the N_F; hence, the assertion is established (we could also have applied the Coherence Lemma (3.1.3) to the sets U_F, regarded as sets of relations with base F). ◁

3.2.6. Again using the ultrafilter axiom:

Given multirelations M, M′ *and* N, *suppose that every multirelation which is a finite restriction of* M *or* M′ *is embeddable in* N. *Then there exists a multirelation* M* *in which* M *and* M′ *are embeddable, whose finite restrictions are embeddable in* N.

▷ Let E and E′ denote the bases of M and M′, and assume them to be disjoint; let D be the base of N. With each ordered pair (F, F′), where F is a finite subset of E and F′ a finite subset of E′, we associate the set $U_{F,F'}$ of mappings ξ of E∪E′ into D whose restrictions to F and F′ are an isomorphism of M | F onto N | ξ(F) and an isomorphism of M′ | F′ onto

$N \mid \xi(F')$, respectively. From our hypotheses it follows that for every F and F′ the set $U_{F,F'}$ is nonempty; furthermore, if $G \supseteq F$ and $G' \supseteq F'$, then $U_{G,G'} \subseteq U_{F,F'}$. The sets $U_{F,F'}$ and their supersets thus constitute a filter over the set of mappings of $E \cup E'$ into D. Now consider a finer ultrafilter.

We now define a "pairing" relation to hold for an element a in E and an element a' in E′ if the mappings ξ such that $\xi(a) = \xi(a')$ constitute an element of the ultrafilter. Only two elements at a time can be paired, one in E and the other in E′. Indeed, pairing is an equivalence relation and, on the other hand, if $a, b \in E$ $(a \neq b)$ the set of all ξ which map $M \mid \{a, b\}$ isomorphically onto $N \mid \{\xi(a), \xi(b)\}$ is an element of the ultrafilter; therefore the set of all ξ for which $\xi(a) \neq \xi(b)$ is an element.

Now take n-ary relations R, R′ and S having the same ranks [positions] in M, M′ and N, respectively, and elements $a_1, \ldots, a_n$ in $E \cup E'$.

Define the relation R^* by the condition: $R^*(a_1, \ldots, a_n) = +$ or $-$ according to whether the mappings ξ such that $S(\xi(a_1), \ldots, \xi(a_n)) = +$ or $-$ constitute an element of the ultrafilter. $R(a_1, \ldots, a_n)$ is obtained if $a_1, \ldots, a_n$ all belong to E, and $R'(a_1, \ldots, a_n)$ if they all belong to E′. Finally, for every finite subset H of $E \cup E'$, there exist mappings ξ constituting an element of the ultrafilter and giving, after identification of the paired elements, an isomorphism of $R^* \mid H$ onto $S \mid \xi(H)$. ◁

3.3. Deducibility of relations

3.3.1. Let R and S be two m-ary relations having the same nonempty base E $(m \geqslant 0)$; we shall say that S *is deducible from* R, writing $R \vdash S$, if every m-tuple over E which gives R the value $+$ also gives S the value $+$. It is clear how one words this condition for the case $m = 0$: we have $(E, -) \vdash (E, +)$, but not conversely.

Example. Starting with any $S(x, y)$, set $R(x, y) = +$ whenever $S(x, y) = +$ and $x \neq y$ as well. In particular, if S is an ordering: $S(x, y) = +$ whenever $x \leqslant y$, then R is the corresponding strict ordering: $R(x, y) = +$ whenever $x < y$.

3.3.2. Let R be an m-ary relation and S an n-ary relation over the same base E. Set $R \vdash S$ whenever every m-tuple over E which gives R the value $+$, when reduced to its first n terms (if $n \leqslant m$) or augmented by $n - m$

arbitrary terms from E (if $n>m$), also gives S the value $+$.

Example. Starting from any $S(x, y)$, set $R(x, y, z)=+$ whenever $S(x, y)=S(y, z)=+$. In particular, if S is an ordering, R is the relation $R(x, y, z)=+$ whenever $x \leqslant y \leqslant z$ in S.

For relations of given base and the same arity, deducibility is an ordering; for relations of arbitrary arity it is a pre-ordering. Indeed, let R be an m-ary relation with base E; for each $n \geqslant m$, the relation

$$R'(x_1, \ldots, x_m, x_{m+1}, \ldots, x_n)=R(x_1, \ldots, x_m)$$

satisfies $R \vdash R'$ and $R' \vdash R$ for any $x_1, \ldots, x_m, x_{m+1}, \ldots, x_n \in E$. We shall say that R and R′ are equideducible and write $R \dashv\vdash R'$; we see that R′ is the only n-ary relation equideducible from R.

Note that a relation over E which is always equal to $+$ is deducible from every relation over E; every relation over E is deducible from a relation which is always equal to $-$.

When either R or S (with nonempty base E) is 0-ary, the above definition applies, if we note that $(E, +)$ takes the value $+$ and $(E, -)$ the value $-$ for the empty sequence. Thus $(E, -) \vdash R \vdash (E, +)$ for any relation R with base E, and $R \dashv\vdash (E, +)$ if and only if R always takes the value $+$. The same holds for $-$.

If the base E is empty, deducibility is no longer transitive: $R \vdash S$ when any sequence that gives R the value $+$, when reduced or arbitrarily augmented, gives S the value $+$. We thus see that all relations with empty base and positive arity are equideducible, each being equideducible from $(\emptyset, +)$ and $(\emptyset, -)$. Finally, we have $(\emptyset, -) \vdash (\emptyset, +)$, but not conversely.

3.3.3. Let R be an m-ary relation and S an n-ary relation over E; let $\bigwedge(R, S)$ denote the relation over E whose arity is the greater of m and n and which is equal to $+$ for those sequences which (when reduced to their first m elements) give the value $+$ to R and (reduced to their first n elements) the value $+$ to S.

Similarly, we define a relation $\bigvee(R, S)$, equal to $+$ for those sequences which give the value $+$ to R *or* S.

Finally, let $\neg R$ denote the relation which always takes the value opposite to that of R. The operators $\bigwedge$, $\bigvee$, and $\neg$ thus defined show that deducibility is a *Boolean pre-lattice* over arbitrary relations with base E, and a *Boolean lattice* [*Boolean algebra*] over the relations with base E and given arity (provided E is not empty).

3.4. Operator, predicate, predicarity and arity; model; selector, + and − operators, identifier, connector, rank-changer, dilator

We shall use the term *operator* for a function $\mathscr{P}$ which, with every multirelation $M=(R_1,\ldots, R_h)$ of given arity $\mu=(m_1,\ldots, m_h)$ and nonempty base, associates a relation $S=\mathscr{P}(M)$ of given arity n having the same base as M. We shall also say that the μ-ary multirelation M is *assignable* to $\mathscr{P}$ and that $\mathscr{P}$ transforms M into S. The value $S(x_1,\ldots, x_n)$ taken by S for a sequence $x_1,\ldots, x_n$ of elements in the base of M will be denoted by $\mathscr{P}(M)(x_1,\ldots, x_n)$.

Examples. The operator which, with every binary relation $R(x, y)$, associates $S(x, y)=R(y, x)$. The operator which, with $R(x, y)$, associates $S(x)=R(x, x)$. The operator, which, with $R(x, y)$, associates the relation $S(x, y)=\neg R(x, y)$ which always takes the opposite value to that of R. The operator which, with every birelation $M=(R, S)$ formed from two binary relations R and S, associates the first of these relations, i.e., R. The quantifiers, introduced below in 3.6, are operators of which we shall make frequent use.

The arity $\mu=(m_1\ldots, m_h)$ of the multirelations M will be called the *predicarity* of the operator $\mathscr{P}$; the arity n of the relations $S=\mathscr{P}(M)$ is the *arity* of $\mathscr{P}$. We shall also say that μ and n are the *arities* of $\mathscr{P}$, or that $\mathscr{P}$ is (μ, n)*-ary* or $(m_1,\ldots, m_h; n)$*-ary*.

In the case of arity 0, the operator $\mathscr{P}$ associates with every M of base E one of the relations (E, +) or (E, −); we shall also say that $\mathscr{P}$ takes the value + or the value −, or that M gives $\mathscr{P}$ the value + or −, or that it does or does not satisfy $\mathscr{P}$. Those M which satisfy $\mathscr{P}$ will be called *models* of $\mathscr{P}$.

In the general case (arbitrary n), we associate with each of the indices $i=1,\ldots, h$ a *predicate*, denoted by ρ_i (one might call this a "relational argument"). Each predicate ρ_i is determined by the index i and the integer m_i, which will be called the *arity* of ρ_i; we shall also say that ρ_i is an m_i*-ary predicate*. Moreover, the operator will sometimes be written $\mathscr{P}(\rho_1,\ldots, \rho_h)$. For each predicate ρ_i one can *substitute* a relation R_i of the same arity, provided only that all the relations $R_1,\ldots, R_h$ have the same base and therefore constitute a multirelation; substitution of $R_1,\ldots, R_h$ for the predicates gives the transformed relation $\mathscr{P}(R_1,\ldots, R_h)$.

We now define some operators which will be used often in the sequel: selector, operators + and −, identifier, and rank-changer, a special case of the latter being the dilator.

3.4.1. Given an arity $\mu=(m_1,\ldots, m_k,\ldots, m_h)$ and an index k, $1\leqslant k\leqslant h$, set $n=m_k$; the μ-ary *selector* of rank k associates with each μ-ary multirelation $\mathrm{M}=(\mathrm{R}_1,\ldots, \mathrm{R}_k,\ldots, \mathrm{R}_h)$ its k-th relation $\mathrm{S}=\mathrm{R}_k$. It thus simply selects the relation of the given rank.

3.4.2. For arities μ and $n\geqslant 0$, we define the *operator* $+^n_\mu$ or $+(x^1\ldots x^n)_\mu$ as the operator which associates with every μ-ary multirelation the n-ary relation having the same base and always equal to +. The operator $-^n_\mu$ is defined in similar fashion. These will be denoted simply by + and −, omitting μ and n, whenever there is no danger of confusion.

3.4.3. Given arities μ and $n\geqslant 2$ and two positive integers $i<j\leqslant n$, we use the term *identifier* $(x^i\equiv x^j)^n_\mu$ for the the operator which associates with every μ-ary multirelation M the n-ary identity relation relative to the indices i and j, i.e., the relation

$$\mathrm{R}(x_1,\ldots, x_i,\ldots, x_j,\ldots, x_n)= + \quad \text{or} \quad -$$

according to whether $x_i=x_j$ or $x_i\neq x_j$ over the base of M. The identifier will be written simply $(x^i\equiv x^j)$, omitting μ and n whenever there is no danger of confusion, particularly when $n=j$. Note that the case $i=j$ gives nothing new; the identifier $(x^i\equiv x^i)$ is simply the operator $+^n_\mu$. We denote by $(x^i\not\equiv x^j)^n_\mu$ the operator which associates with every μ-ary multirelation M the n-ary inequality relation over the base of M.

3.4.4. Let $\mu=(m_1,\ldots, m_h)$ be a sequence of integers; let n be at least equal to their maximum and let α be an h-ary connection. We shall use the term *connector* α^n_μ for the operator which associates with every μ-ary multirelation $\mathrm{M}=(\mathrm{R}_1,\ldots, \mathrm{R}_h)$ the n-ary relation S defined by

$$\mathrm{S}(x_1,\ldots, x_n)=\alpha(\mathrm{R}_1(x_1,\ldots, x_m),\ldots, R_h(x_1,\ldots, x_{m_h}))$$

for arbitrary $x_1,\ldots, x_n$ in the base of M. The connector will be denoted simply by the connection α, omitting μ and n whenever there is no danger of confusion.

Examples. The connector $\neg_n^n$ transforms every n-ary relation R into the relation $\neg R_n^n$, or simply $\neg R$, which always takes the value opposite to that of R. The connector $\bigwedge_{m_1, m_2}^n$ transforms every multirelation (R_1, R_2) in which R_1 is m_1-ary and R_2 is m_2-ary, into $S = \bigwedge_{m_1, m_2}^n (R_1, R_2)$, or simply $\bigwedge(R_1, R_2)$, which equals + for those n-tuples ($n \geqslant$ maximum of m_1 and m_2) giving the value + to both R_1 and R_2. $\bigvee_{m_1, m_2}^n$ is defined in similar fashion, with $\bigvee(R_1, R_2)$ equal to + whenever either R_1 or R_2 take the value +. Special cases are the operations $\neg$, $\bigwedge$, and $\bigvee$ already considered in 3.3.3.

The definition of the connector $\Rightarrow_{m_1, m_2}^n$ is similar to the preceding definitions. One sees that $R \vdash S$ (see 3.3), where R is m-ary and S n-ary, if and only if the relation $\Rightarrow(R, S)$ is always equal to +.

3.4.5. Let m and n be two positive integers and f a mapping of $\{1, \ldots, m\}$ into $\{1, \ldots, n\}$. We shall use the term *rank-changer* f_m^n for the operator which associates with every m-ary relation R the n-ary relation S, with the same base, such that

$$S(x_1, \ldots, x_n) = R(x_{f(1)}, \ldots, x_{f(m)})$$

for any $x_1, \ldots, x_n$ in the base. The rank-changer will also be written $(\rho x^{f(1)} \ldots x^{f(m)})^n$. *We shall omit the index n whenever n is the maximum of* $f(1), \ldots, f(m)$; in the general case we only have $n \geqslant$ the maximum of $f(1), \ldots, f(m)$. We also define the rank-changer for $m=0$, $n \geqslant 0$: then, for any n-tuple, $\rho = f_0^n R$ takes the value (+ or −) of R. Note that $f_0^n R$ is defined even when the base is empty.

Special cases. The identity operator, for which $S = R$; this is simply the changer $(\rho x^1 \ldots x^m)^m$.

The *dilator*, which associates with every m-ary relation R the n-ary relation ($n \geqslant m$)

$$S(x_1, \ldots, x_m, x_{m+1}, \ldots, x_n) = R(x_1, \ldots, x_m),$$

which is simply the changer $(\rho x^1 \ldots x^m)^n$. Note that we have already encountered S, as a relation equideducible from R (see 3.3). The changer f_n^m, where $f(i) = i$ for $i = 1, \ldots, m$ and, say, $f(i) = 1$ for $i = m+1, \ldots, n$ (also written as $(\rho x^1 \ldots x^m x^1 \ldots x^1)^m$) converts S back into R (with $m \geqslant 1$).

Among the examples already considered at the beginning of 3.4 are: the changer ρxx, which transforms each $R(x, y)$ into $S(x) = R(x, x)$ (the theoretical notation is $(\rho x^1 x^1)$, or f_2^1, with $f(1)=f(2)=1$), and the changer ρyx, which transforms each $R(x, y)$ into $S(x, y)=R(y, x)$ (the theoretical notation is $(\rho x^2 x^1)^2$, or f_2^2 with $f(1)=2$ and $f(2)=1$).

3.4.6. When the base is empty, as we have seen, there exists only one n-ary relation (for positive n), and so all the above operators transform every μ-ary multirelation with empty base into the unique relation R with empty base; in particular, $\neg_n^n R=R$.

When the arity is 0 and the base E arbitrary (possibly empty), the operator $+_\mu^0$ yields the relation (E, +), while $-_\mu^0$ yields (E, −). Every connector $\alpha_{0,\ldots,0}^0$ operates on the relations (E, +) and (E, −) in the same way as the connection α operates on the values + and −.

3.5. Deducibility of Operators; Composite-Operator; Image and Inverse Image of a Class

3.5.1. Let $\mathscr{P}$ and $\mathscr{Q}$ be two operators of the same predicarity μ and arbitrary arities (which may be different). We shall say that $\mathscr{Q}$ is deducible from $\mathscr{P}$, writing $\mathscr{P} \vdash \mathscr{Q}$, if $\mathscr{P}(M) \vdash \mathscr{Q}(M)$ for every μ-ary multirelation M. In other words, if p and q are the arities of $\mathscr{P}$ and $\mathscr{Q}$, whenever $\mathscr{P}(M)(x_1, \ldots, x_p)=+$ implies $\mathscr{Q}(M)(x_1, \ldots, x_q)=+$ for every μ-ary multirelation M and every sequence $x_1, x_2, \ldots$ in which the number of terms equals the greater of p and q.

Deducibility $\vdash$ is a pre-ordering on operators of given predicarity and an ordering when restricted to operators of given arity. A + operator is deducible from every other operator, and every other operator is deducible from a − operator.

3.5.2. Let $\mathscr{P}$ be an $(m_1, \ldots, m_h; n)$-ary operator, and $\mathscr{Q}_1, \ldots, \mathscr{Q}_h$ operators with the same predicarity ν and arities $m_1, \ldots, m_h$. Then the *composite operator* $\mathscr{P}(\mathscr{Q}_1, \ldots, \mathscr{Q}_h)$ is the operator which transforms every ν-ary multirelation N into $\mathscr{P}(\mathscr{Q}_1(N), \ldots, \mathscr{Q}_h(N))$.

Example. Let $\mathscr{P}$ transform each multirelation $(R(x, y), S(x))$ into the relation $T(x, y)=+$ whenever $R(x, y)=S(x)=+$. Let $\mathscr{Q}$ transform each $R(x, y)$ into $R(y, x)$, and let $\mathscr{R}$ transform each $R(x, y)$ into the unary relation $S(x)$ which equals + whenever $R(x, z)=+$ for arbitrary z. Then $\mathscr{P}(\mathscr{Q}, \mathscr{R})$ will transform every relation $R(x, y)$ into $T(x, y)=+$ whenever

$R(y, x) = +$ and in addition $R(x, z) = +$ for arbitrary z.

Given two operators, $\mathscr{P}$ of arity p and $\mathscr{Q}$ of arity q, the composite operators $\bigwedge_{p,q}(\mathscr{P}, \mathscr{Q})$ and $\bigvee_{p,q}(\mathscr{P}, \mathscr{Q})$ and the composite operator $\neg_p \mathscr{P}$ show that deducibility of operators is a Boolean pre-lattice (a Boolean lattice [Boolean algebra] over operators of given arity; see 3.3). $\mathscr{P} \vdash \mathscr{Q}$ if and only if the composite operator $\Rightarrow_{p,q}(\mathscr{P}, \mathscr{Q})$ transforms every multirelation assignable to it into a relation which is always equal to $+$, or, briefly, if and only if it is a $+$ operator (see 3.4.2).

3.5.3. Let f be a mapping of $\{1,..., m\}$ into $\{1,..., m'\}$ and g a mapping of $\{1,..., m'\}$ into $\{1,..., m''\}$; then the *composite of* $f_m^{m'}$ *and* $g_{m'}^{m''}$ *is* $(gf)_m^{m''}$.

Indeed, if R is an m-ary relation, set

$$S = f_m^{m'} R \quad \text{and} \quad T = g_{m'}^{m''} S;$$

then

$$T(x_1, ..., x_{m''}) = S(x_{g(1)}, ..., x_{g(m')})$$

for any $x_1, ..., x_{m''}$ in the base, and

$$S(x_1, ..., x_{m'}) = R(x_{f(1)}, ..., x_{f(m)})$$

for any $x_1, ..., x_{m'}$; in particular, this holds for $x_{g(1)}, ..., x_{g(m')}$, giving

$$S(x_{g(1)}, ..., x_{g(m')}) = R(x_{gf(1)}, ..., x_{gf(m)}).$$

3.5.4. *Let* α *be an h-ary connection,* f *a mapping of* $\{1,..., m\}$ *into* $\{1,..., n\}$, *and* $m_1, ..., m_h$ *a sequence of integers whose maximum is m. Let* f_i *denote the restriction of* f *to* $\{1,..., m_i\}$, *and* s_i *the selector which transforms every* $(m_1, ..., m_h)$*-ary multirelation into its i-th relation.*

Then the composite operator $f_m^n \alpha_{m_1, ..., m_h}^m$ *is equal to the composite*

$$\alpha_{n, ..., n}^n((f_1)_{m_1}^n s_1, ..., (f_h)_{m_h}^n s_h).$$

Indeed, a multirelation $(R_1, ..., R_h)$ of arity $(m_1, ..., m_h)$ is transformed by either of the operators under consideration into the operator S such that

$$S(x_1, ..., x_n) = \alpha(R_1(x_{f(1)}, ..., x_{f(m_1)}), ..., R_h(x_{f(1)}, ..., x_{f(m_h)})).$$

3.5.5. Let R and S be two relations with the same arity m and the same base, and f a mapping of $\{1,\ldots, m\}$ into $\{1,\ldots, n\}$; then *if* $R \vdash S$, *it follows that* $f_m^n R \vdash f_m^n S$. Indeed, from $R(x_{f(1)},\ldots, x_{f(m)}) = +$ and $R \vdash S$, it follows that $S(x_{f(1)},\ldots, x_{f(m)}) = +$.

In general, given R and S with arbitrary arities m and m' and the same base, let f' be a mapping of $\{1,\ldots, m'\}$ into $\{1,\ldots, n'\}$ which is either an extension of f (if $m' \geqslant m$) or a restriction of f (if $m' \leqslant m$); then *if* $R \vdash S$, *it follows that* $f_m^n R \vdash f_{m'}^{\prime n'} S$.

3.5.6. Let $\mathscr{A}$ be a class of μ-ary multirelations and $\mathscr{P}$ an operator of predicarity μ; the *image of* $\mathscr{A}$ *under* $\mathscr{P}$ will be the class of all relations $\mathscr{P}(M)$ for $M \in \mathscr{A}$. Taking a finite sequence of operators $\mathscr{P}_1,\ldots, \mathscr{P}_h$, we can define the more general *image of* $\mathscr{A}$ *under* $\mathscr{P}_1,\ldots, \mathscr{P}_h$, where $M \in \mathscr{A}$.

Example. Let $\mathscr{A}$ be the class of chains or total orderings which have no maximal element, and thus have infinite base; let $\mathscr{P}$ be the operator which associates with each binary relation the unary relation with the same base which is always true; we get the class of unary relations which are always true and have infinite base.

Let $\mathscr{A}$ be a class of ν-ary multirelations, where $\nu = n_1,\ldots, n_h$, and $\mathscr{P}_1,\ldots, \mathscr{P}_h$ operators having the same predicarity μ and arities $n_1,\ldots, n_h$, then the *inverse image of* $\mathscr{A}$ *under* $\mathscr{P}_1,\ldots, \mathscr{P}_h$ is the class of μ-ary multirelations M such that $(\mathscr{P}_1(M),\ldots, \mathscr{P}_h(M)) \in \mathscr{A}$.

Example. Let $\mathscr{P}$ be the same operator as in the preceding example and let $\mathscr{A}$ be the class of infinite unary relations, then the inverse image is the class of infinite binary relations.

3.6. Active or Inactive Index; Quantifier; Composite Quantifier

3.6.1. Let R be an m-ary relation; an integer $i \leqslant m$ will be called an *inactive index* in R if the value $R(x_1,\ldots, x_i,\ldots, x_m)$ is the same for any two m-tuples over the base which differ only in their i-th term. Otherwise the index will be said to be *active*.

Example. If $R(x_1, x_2)$ is always equal to $+$, both indices 1 and 2 are inactive. If $R(x_1, x_2)$ takes the value $+$ over the natural numbers when x_1 is even and the value $-$ when x_1 is odd, irrespective of the value of x_2, then the index 1 is active and the index 2 is inactive. If $R(x_1, x_2)$ is the chain of integers ($+$ when $x_1 \leqslant x_2$) then the indices 1 and 2 are active.

Let $\mathscr{P}$ be an operator, n its arity; an integer $\leqslant n$ is said to be an *inactive index* in $\mathscr{P}$ if it is inactive in the relation $\mathscr{P}(\mathrm{M})$ for any multirelation M assignable to $\mathscr{P}$. For example, 1 is inactive in the rank changer ρx^2 or in the identifier $x^2 \equiv x^3$. The index i is *active* otherwise, i.e., if there exists at least one M such that i is active in $\mathscr{P}(\mathrm{M})$.

3.6.2. Given positive integers m and $i \leqslant m$, we define the *universal* (m, m)*-ary quantifier* of index i, denoted by $\underset{i}{\forall}{}^m_m$, as the operator which associates with each m-ary relation R the m-ary relation S with the same base E such that

$$\mathrm{S}(x_1, \ldots, x_i, \ldots, x_m) = + \quad \text{if} \quad R(x_1, \ldots, x', \ldots, x_m) = +$$

for any $x' \in \mathrm{E}$ substituted for x_i. The other terms $x_1, \ldots, x_{i-1}, x_{i+1}, \ldots, x_m$ remain fixed. Otherwise we have

$$\mathrm{S}(x_1, \ldots, x_i, \ldots, x_m) = - .$$

The definition of the *existential quantifier* $\underset{i}{\exists}{}^m_m$ is the same, with the words "for any" replaced by "for at least one".

When R is the m-ary relation with empty base, we have

$$\underset{i}{\forall}{}^m_m \mathrm{R} = \underset{i}{\exists}{}^m_m \mathrm{R} = \mathrm{R} .$$

It is sometimes convenient to define the quantifiers $\underset{i}{\forall}{}^m_m$ and $\underset{i}{\exists}{}^m_m$ for $i > m$; these are simply identity operators (leaving every m-ary relation unchanged).

For the case $i = m$, besides the (m, m)-ary quantifiers $\underset{m}{\forall}{}^m_m$ and $\underset{m}{\exists}{}^m_m$, we also define the $(m, m-1)$-ary quantifiers $\underset{m}{\forall}{}^{m-1}_m$ and $\underset{m}{\exists}{}^{m-1}_m$; they transform R into the relation S such that

$$\mathrm{S}(x_1, \ldots, x_{m-1}) = + \quad \text{if} \quad \mathrm{R}(x_1, \ldots, x_{m-1}, x') = +$$

for any $x' \in \mathrm{E}$ substituted for x_m (or for at least one x' substituted for x_m). The relations $\underset{m}{\forall}{}^m_m \mathrm{R}$ and $\underset{m}{\forall}{}^{m-1}_m \mathrm{R}$ take the same values whenever $x_1, \ldots, x_{m-1}$ are given; they are therefore equideducible. In questions involving deducibility there is in general no need to specify the arity of the quantifier.

Example. Let $\mathrm{R}(x_1, x_2)$ be the chain (total ordering) of the natural numbers, equal to $+$ whenever $x_1 \leqslant x_2$. Then $\underset{1}{\forall}{}^2_2 \mathrm{R}$ is the binary relation on the natural numbers which is always $-$; the relation $\underset{2}{\forall}{}_2^2\mathrm{R}$ is binary, equal to $+$ for every pair $(0, x_2)$ and to $-$ for every pair (x_1, x_2) such that $x_1 \neq 0$; the relation $\underset{2}{\forall}{}_2^1\mathrm{R}$ is unary, equal to $+$ for $x_1 = 0$ and to $-$ for $x_1 \neq 0$. The relations $\underset{1}{\exists}{}_2^2\mathrm{R}$, $\underset{2}{\exists}{}_2^2\mathrm{R}$, and $\underset{2}{\exists}{}_2^1\mathrm{R}$ always take the value $+$.

When using the notation $\mathrm{R}(x, y, z, \ldots)$ we shall write the quantifiers $\underset{1}{\forall}, \underset{2}{\forall}, \underset{3}{\forall}, \ldots$ as $\underset{x}{\forall}, \underset{y}{\forall}, \underset{z}{\forall}, \ldots$; an analogous convention holds for $\exists$.

We must also define *two quantifiers* $\underset{1}{\forall}{}_1^0$; one of these transforms the unary relation with empty base into $(\emptyset, +)$, the other transforms it into $(\emptyset, -)$. The same distinction is necessary for $\exists_1^0$ and the composite quantifiers $\underset{1,2,\ldots,m}{\forall_m^0}$ and $\underset{1,2,\ldots,m}{\exists_m^0}$ (see below, 3.6.4).

3.6.3. In the rest of this chapter, we shall assume that the base is not empty.

In the relations $\underset{i}{\forall}\mathrm{R}$ and $\underset{i}{\exists}\mathrm{R}$, the index i is inactive. If j is inactive in R then it is also inactive in $\underset{i}{\forall}\mathrm{R}$ and in $\underset{i}{\exists}\mathrm{R}$.

For every R we have $\underset{i}{\forall}\mathrm{R} \vdash \mathrm{R} \vdash \underset{i}{\exists}\mathrm{R}$. If the index i is inactive in R, then $\underset{i}{\forall}\mathrm{R} -\!\vdash \underset{i}{\exists}\mathrm{R} -\!\vdash \mathrm{R}$; conversely, if $\underset{i}{\forall}\mathrm{R} -\!\vdash \mathrm{R}$ or $\underset{i}{\exists}\mathrm{R} -\!\vdash \mathrm{R}$, then i is inactive in R.

If $\mathrm{R} \vdash \mathrm{S}$, then $\underset{i}{\forall}\mathrm{R} \vdash \underset{i}{\forall}\mathrm{S}$ and $\underset{i}{\exists}\mathrm{R} \vdash \underset{i}{\exists}\mathrm{S}$.

If $\mathrm{R} \vdash \mathrm{S}$ and the index i is inactive in R, then $\mathrm{R} \vdash \forall\mathrm{S}$; if the index j is inactive in S, then $\underset{j}{\exists}\mathrm{R} \vdash \mathrm{S}$. If both i is inactive in R and j is inactive in S, then

$$\mathrm{R} \vdash \underset{j}{\exists}\mathrm{R} \vdash \underset{i}{\forall}\mathrm{S} \vdash \mathrm{S}$$

and i and j are both inactive in $\underset{j}{\exists}\mathrm{R}$ and in $\underset{i}{\forall}\mathrm{S}$.

3.6.4. We define the *composite quantifier*, denoted by

$$\mathscr{P} = \underset{ij\ldots}{\forall}\ \underset{i'j'\ldots}{\exists}\ \underset{i''j''\ldots}{\forall_m^m},$$

as the composite of the simple quantifiers

$$\begin{pmatrix}\forall^m_m\\ i\end{pmatrix}\begin{pmatrix}\forall^m_m\\ j\end{pmatrix}\cdots\begin{pmatrix}\exists^m_m\\ i'\end{pmatrix}\begin{pmatrix}\exists^m_m\\ j'\end{pmatrix}\cdots\begin{pmatrix}\forall^m_m\\ i''\end{pmatrix}\begin{pmatrix}\forall^m_m\\ j''\end{pmatrix}\cdots,$$

of arities (m, m). The integers $i, j, \ldots, i', j', \ldots$ are called the *indices* of $\mathscr{P}$.

For every m-ary relation R, the indices of $\mathscr{P}$ are inactive in $\mathscr{P}$R.

Let $0 \leqslant n < m$; if each of the integers $n+1, n+2, \ldots, m$ appears among the indices of $\mathscr{P}$, they are inactive in $\mathscr{P}$R. We now define the composite quantifier $\mathscr{P}' = \underset{ij\ldots i'j'\ldots i''j''\ldots}{\forall\ \exists\ \forall^n_m}$ of arities (m, n) by the equality

$$(\mathscr{P}'\mathrm{R})(x_1, \ldots, x_n) = (\mathscr{P}R)(x_1, \ldots, x_n, x_{n+1}, \ldots, x_m)$$

for every m-ary relation R and every $x_1, \ldots, x_n$. We can also define $\mathscr{P}'$ as the composite $f^n_m\mathscr{P}$, where f^n_m is the rank-changer defined as $f(i) = i$ for $i = 1, \ldots, n$ and $f(n+1), \ldots, f(m)$ are arbitrary numbers $\leqslant n$.

The simple quantifiers $\forall$ commute, [and so do the quantifiers $\exists$]:

$$\underset{i}{\forall}\underset{j}{\forall}\mathrm{R} \dashv\vdash \underset{j}{\forall}\underset{i}{\forall}\mathrm{R} \quad \text{and} \quad \underset{i}{\exists}\underset{j}{\exists}\mathrm{R} \dashv\vdash \underset{j}{\exists}\underset{i}{\exists}\mathrm{R};$$

in the composite quantifier $\mathscr{P}'$, we can therefore change the order of $i, j, \ldots,$ of $i', j', \ldots,$ of $i'', j'', \ldots$.

The quantifiers $\underset{i}{\forall}$ and $\underset{j}{\exists}$ do not commute. To see this, consider the ordering $\mathrm{R}(x_1, x_2)$ of the natural numbers; since the relation $\underset{1}{\forall}\mathrm{R}$ is always equal to $-$, the same holds for $\underset{2}{\exists}\underset{1}{\forall}\mathrm{R}$; since the relation $\underset{2}{\exists}\mathrm{R}$ is always equal to $+$, the same holds for $\underset{1}{\forall}\underset{2}{\exists}\mathrm{R}$.

If the indices of $\mathscr{P}$ are inactive in R, then $\mathscr{P}\mathrm{R} \dashv\vdash \mathrm{R}$, and vice versa.

Consequently, if every index of a quantifier $\mathscr{Q}$ already appears in $\mathscr{P}$, then $\mathscr{Q}\mathscr{P} \dashv\vdash \mathscr{P}$. In particular, if $\mathscr{P}$ is (m, m)-ary so that the composite operator is defined then $\mathscr{P}\mathscr{P} \dashv\vdash \mathscr{P}$.

If $\mathrm{R} \vdash \mathrm{S}$, then $\mathscr{P}\mathrm{R} \vdash \mathscr{P}\mathrm{S}$.

If $\mathrm{R} \vdash \mathrm{S}$ and the indices of $\mathscr{P}$ are inactive in R, then $\mathrm{R} \vdash \mathscr{P}\mathrm{S}$; if the indices of $\mathscr{P}$ are inactive in S, then $\mathscr{P}\mathrm{R} \vdash \mathrm{S}$.

3.7. Composite of Quantifier and Connector

3.7.1. The following equideductions hold for all relations R and S:

$$\neg\underset{i}{\forall}\mathrm{R} \dashv\vdash \underset{i}{\exists}\neg\mathrm{R} \quad \text{and} \quad \neg\underset{i}{\exists}\mathrm{R} \dashv\vdash \underset{i}{\forall}\neg\mathrm{R};$$

$$\underset{i}{\forall}(R\bigwedge S) \dashv\vdash \underset{i}{\forall}R\bigwedge\underset{i}{\forall}S;$$
$$\underset{i}{\exists}(R\bigvee S) \dashv\vdash \underset{i}{\exists}R\bigvee\underset{i}{\exists}S.$$

The following deductions hold:

$$\underset{i}{\forall}R\bigvee\underset{i}{\forall}S \vdash \underset{i}{\forall}(R\bigvee S);$$
$$\underset{i}{\exists}(R\bigwedge S) \vdash \underset{i}{\exists}R\bigwedge\underset{i}{\exists}S.$$

Indeed,

$$R\vdash R\bigvee S, \quad \text{whence} \quad \underset{i}{\forall}R\vdash\underset{i}{\forall}(R\bigvee S)$$

and, similarly, $\underset{i}{\forall}S\vdash\underset{i}{\forall}(R\bigvee S)$; the first deduction then follows. The proof of the second is similar.

Note that the preceding deductions are in general not equideductions. Thus, let R and S be relations on $\{0, 1\}$ such that $R(0)=+$, $R(1)=-$, $S(0)=-$ and $S(1)=+$. Then $R\bigvee S$ is always $+$, and so $\underset{i}{\forall}(R\bigvee S)$ is always $+$, while $\underset{1}{\forall}R$, $\underset{1}{\forall}S$ and $\underset{1}{\forall}R\bigvee\underset{1}{\forall}S$ are always $-$.

3.7.2. (1) *If the index i is inactive in* R, the following equideductions hold:

$$\underset{i}{\forall}(R\bigwedge S)\dashv\vdash R\bigwedge\underset{i}{\forall}S; \qquad \underset{i}{\forall}(R\bigvee S)\dashv\vdash R\bigvee\underset{i}{\forall}S;$$
$$\underset{i}{\exists}(R\bigwedge S)\dashv\vdash R\bigwedge\underset{i}{\exists}S; \qquad \underset{i}{\exists}(R\bigvee S)\dashv\vdash R\bigvee\underset{i}{\exists}S.$$

▷ The first equideduction follows immediately from

$$\underset{i}{\forall}(R\bigwedge S)\dashv\vdash\underset{i}{\forall}R\bigwedge\underset{i}{\forall}S \quad \text{and} \quad \underset{i}{\forall}R\dashv\vdash R.$$

The proof of the fourth equideduction is the same.

For the second equideduction we start from

$$\underset{i}{\forall}R\bigvee\underset{i}{\forall}S\vdash\underset{i}{\forall}(R\bigvee S) \quad \text{and} \quad \underset{i}{\forall}R\dashv\vdash R,$$

which gives $R\bigvee\underset{i}{\forall}S\vdash\underset{i}{\forall}(R\bigvee S)$. On the other hand, suppose that

$$\left(\underset{i}{\forall}(R\bigvee S)\right)(x_1,\ldots, x_i,\ldots, x_n)=+;$$

then $(R \vee S)(x_1, \ldots, x', \ldots, x_n) = +$ for any value x' substituted for x_i; therefore, either $R(x_1, \ldots, x_i, \ldots, x_n) = +$, and $R \vee \underset{i}{\forall} S$ is *a fortiori* $+$, or $R(x_1, \ldots, x_i, \ldots, x_n) = -$ and, since the index i is inactive, $R(x_1, \ldots, x', \ldots, x_n) = -$ for every value x' substituted for x_i, in which case $S(x_1, \ldots, x', \ldots, x_n) = +$ for every x', whence

$$\left(\underset{i}{\forall} S\right)(x_1, \ldots, x_i, \ldots, x_n) = +,$$

and so $R \vee \underset{i}{\forall} S$ is $+$. It follows that $\underset{i}{\forall}(R \vee S) \vdash R \vee \underset{i}{\forall} S$. The proof of the third equideduction is the same. ◁

(2) In general, let $\mathscr{P}$ be a composite quantifier. *Then, if the indices of* $\mathscr{P}$ *are inactive* in R, we have

$$\mathscr{P}(R \wedge S) \dashv\vdash R \wedge \mathscr{P} S; \qquad \mathscr{P}(R \vee S) \dashv\vdash R \vee \mathscr{P} S.$$

It suffices to apply one of the preceding equideductions in succession to each of the quantifiers in $\mathscr{P}$. For example, if i and j are inactive in R, we get, successively,

$$\underset{j}{\exists}\underset{i}{\forall}(R \wedge S) \dashv\vdash \underset{j}{\exists}\left(R \wedge \underset{i}{\forall} S\right) \dashv\vdash R \wedge \underset{j}{\exists}\underset{j}{\forall} S.$$

3.7.3. *For all relations* R *and* S,

$$\underset{i}{\forall}\left(\underset{i}{\forall} R \wedge S\right) \dashv\vdash \underset{i}{\forall} R \wedge \underset{i}{\forall} S; \qquad \underset{i}{\forall}\left(\underset{i}{\exists} R \wedge S\right) \dashv\vdash \underset{i}{\exists} R \wedge \underset{i}{\forall} S;$$

$$\underset{i}{\exists}\left(\underset{i}{\forall} R \wedge S\right) \dashv\vdash \underset{i}{\forall} R \wedge \underset{i}{\exists} S; \qquad \underset{i}{\exists}\left(\underset{i}{\exists} R \wedge S\right) \dashv\vdash \underset{i}{\exists} R \wedge \underset{i}{\exists} S$$

and, replacing $\wedge$ by $\vee$, we get four other equideductions. This follows from 3.7.2 (1).

In general, let $\mathscr{P}$ and $\mathscr{Q}$ be quantifiers such that *every index* $\mathscr{P}$ *is an index of* $\mathscr{Q}$. Then, *for all relations* R *and* S,

$$\mathscr{P}(\mathscr{Q} R \wedge S) \dashv\vdash \mathscr{Q} R \wedge \mathscr{P} S; \qquad \mathscr{P}(\mathscr{Q} R \vee S) \dashv\vdash \mathscr{Q} R \vee \mathscr{P} S.$$

For every composite quantifier $\mathscr{P}$ *and all relations* R *and* S,

$$\mathscr{P}(\mathscr{P} R \wedge S) \dashv\vdash \mathscr{P} R \wedge \mathscr{P} S; \qquad \mathscr{P}(\mathscr{P} R \vee S) \dashv\vdash \mathscr{P} R \vee \mathscr{P} S;$$

these follow from 3.7.2 (2).

3.7.4. *For all* $\mathscr{P}$ *and* $\mathscr{Q}$ *and all* R,

$$\mathscr{P}(\mathscr{P}\mathrm{R}\wedge\mathrm{R})-\!\vdash\mathscr{P}(\mathscr{P}\mathrm{R}\vee\mathrm{R})-\!\vdash\mathscr{P}\mathrm{R},$$
$$\mathscr{Q}\mathscr{P}(\mathscr{P}\mathrm{R}\wedge\mathrm{R})-\!\vdash\mathscr{Q}\mathscr{P}(\mathscr{P}\mathrm{R}\vee\mathrm{R})-\!\vdash\mathscr{Q}\mathscr{P}\mathrm{R}.$$

Note that, on the other hand, $\mathscr{Q}\mathscr{P}(\mathscr{Q}\mathrm{R}\vee\mathrm{R})$ is in general not equideducible from $\mathscr{Q}\mathscr{P}\mathrm{R}$. For example, let I be the identity on a set of several elements; then, taking $\underset{2}{\forall}$ as $\mathscr{P}$ and $\underset{1}{\exists}$ as $\mathscr{Q}$, we see that $\underset{1}{\exists}\underset{2}{\forall}((\underset{1}{\exists}\mathrm{I})\vee\mathrm{I})$ is $+$, for the simple reason that $(\underset{1}{\exists}\mathrm{I})\,(x_1, x_2)$ is always equal to $+$ (there exists a value x_1 identical to x_2), whereas $(\underset{1}{\exists}\underset{2}{\forall}\mathrm{I})\,(x_1, x_2)$ is always $-$ (there does not exist an x_1 identical to every x_2).

3.7.5. *For all* $\mathscr{P}$ *and all* R *and* S,

$$\mathscr{P}(\mathrm{R}\wedge\mathscr{P}\mathrm{R}\wedge\mathrm{S})-\!\vdash\mathscr{P}(\mathrm{R}\wedge\mathrm{S}),$$

and the same holds with $\wedge$ replaced by $\vee$.

▷ We have $\mathrm{R}\wedge\mathscr{P}\mathrm{R}\wedge\mathrm{S}\vdash\mathrm{R}\wedge\mathrm{S}$, and so can deduce the second member of the required equivalence from its first member. Conversely, we have $\mathrm{R}\wedge\mathrm{S}\vdash\mathrm{R}$, whence $\mathscr{P}(\mathrm{R}\wedge\mathrm{S})\vdash\mathscr{P}\mathrm{R}$, so that $\mathrm{R}\wedge\mathrm{S}\wedge\mathscr{P}(\mathrm{R}\wedge\mathrm{S})\vdash\mathrm{R}\wedge\mathscr{P}\mathrm{R}\wedge\mathrm{S}$; replacing R by $\mathrm{R}\wedge\mathrm{S}$ in the equivalence $\mathscr{P}(\mathscr{P}\mathrm{R}\wedge\mathrm{R})-\!\vdash\mathscr{P}\mathrm{R}$ of 3.7.4, we get

$$\mathscr{P}(\mathrm{R}\wedge\mathrm{S})-\!\vdash\mathscr{P}(\mathrm{R}\wedge\mathrm{S}\wedge\mathscr{P}(\mathrm{R}\wedge\mathrm{S}))\vdash\mathscr{P}(\mathrm{R}\wedge\mathscr{P}\mathrm{R}\wedge\mathrm{S}).\ \triangleleft$$

3.8. Composite of Quantifier and Rank-Changer

3.8.1. *Let f be a mapping of* $\{1,\ldots, m\}$ *into* $\{1,\ldots, n\}$ *which takes the value* $f(i)$ *only once*; then, for every m-ary relation R, the rank-changer f_m^n defined in 3.4.5 and the quantifiers $\underset{i}{\forall}$ and $\underset{i}{\exists}$ $(1\leqslant i\leqslant m)$ satisfy the identities

$$f_m^n\underset{i}{\forall_m^m}\mathrm{R}=\underset{f(i)}{\forall_n^n}f_m^n\mathrm{R};\qquad f_m^n\underset{i}{\exists_m^m}\mathrm{R}=\underset{f(i)}{\exists_n^n}f_m^n\mathrm{R}.$$

▷ Set $\mathrm{S}=f_m^n\mathrm{R}$ and $\mathrm{S}'=\underset{f(i)}{\forall_n^n}\mathrm{S}$; then

$$\mathrm{S}(x_1,\ldots, x_n)=\mathrm{R}(x_{f(1)},\ldots, x_{f(i)},\ldots, x_{f(m)})$$

for any $x_1,\ldots, x_n$ and $\mathrm{S}'(x_1,\ldots, x_n)=+$ if and only if $\mathrm{S}(x_1,\ldots, x',\ldots, x_n)$

$=+$ for any x' substituted for $x_{f(i)}$; in other words, if and only if

$$\mathrm{R}(x_{f(1)},\ldots,x',\ldots,x_{f(m)})=+$$

for every common value x' substituted simultaneously for $x_{f(i)}$ and for all $x_{f(j)}$ such that $f(j)=f(i)$ in the sequence $(x_{f(1)},\ldots,x_{f(i)},\ldots,x_{f(j)},\ldots,x_{f(m)})$. By assumption, we need consider only the i-th term $x_{f(i)}$ in the sequence.

On the other hand, set $\mathrm{R}'=\underset{i}{\forall}{}^m_m\mathrm{R}$ and $\mathrm{S}''=f^n_m\mathrm{R}'$; then

$$\mathrm{S}''(x_1,\ldots,x_n)=\mathrm{R}'(x_{f(1)},\ldots,x_{f(i)},\ldots,x_{f(m)})$$

for every $x_1,\ldots,x_n$, so that $\mathrm{S}''(x_1,\ldots,x_n)=+$ if and only if $\mathrm{R}(x_{f(1)},\ldots,x',\ldots,x_{f(m)})=+$ for every x' substituted for the i-th term alone, $x_{f(i)}$, of the sequence $(x_{f(1)},\ldots,x_{f(i)},\ldots,x_{f(m)})$. The required identity in $\forall$ is thus established. The proof is the same for the identity in $\exists$. $\lhd$

In general, let $\mathscr{P}^m_m$ be an (m,m)-ary composite quantifier with indices $i,j,\ldots$, and *f a mapping of* $\{1,\ldots,m\}$ *into* $\{1,\ldots,n\}$ *which takes each value* $f(i),f(j),\ldots$ *only once.* Let $\mathscr{P}_{f^n_n}$ denote the (n,n)-ary quantifier obtained from $\mathscr{P}$ by replacing $i,j,\ldots$ by $f(i)$, $f(j),\ldots$. Then $f^n_m\mathscr{P}^m_m\mathrm{R}=\mathscr{P}_{f^n_n}\,f^n_m\mathrm{R}$ *for every m-ary relation* R.

Special cases. (1) The mapping f is injective on $\{1,\ldots,m\}$ (and so $n\geqslant m$).

(2) f is the identity on the indices $i,j,\ldots$ of the quantifier $\mathscr{P}$ and takes each value $i,j,\ldots$only once; then $f^n_m\mathscr{P}^m_m=\mathscr{P}^n_nf^n_m$, where $\mathscr{P}^n_n$ has the same indices and the same simple quantifiers ($\forall$ or $\exists$) as $\mathscr{P}^m_m$ and differs from it only in its arities.

3.8.2. When f takes the same value $f(i)$ several times, the preceding conclusion may be false.

Example. Let R be the chain of natural numbers and let $m=2$, $n=1$, and $f(1)=f(2)=1$. Then $\underset{1}{\forall}{}^2_2\,\mathrm{R}$ is the binary relation on the integers which is always equal to $-$, since no a_2 satisfies $a_1\leqslant a_2$ for all a_1; hence $f^1_2\underset{1}{\forall}{}^2_2\mathrm{R}$ is the unary relation which is always equal to $-$. On the other hand, $f^1_2\mathrm{R}$ is the relation S such that $\mathrm{S}(a_1)=\mathrm{R}(a_1,a_1)=+$ for all a_1; it therefore follows that $\underset{1}{\forall}{}^1_1f^1_2\mathrm{R}$ is the unary relation which is always equal to $+$.

3.8.3. *Let $i,j,\ldots$ be integers and f a mapping of* $1,\ldots,m$ *into* $1,\ldots,n$ *which is* *the identity on the integers other than i, j,...; then*

$$\underset{ij\ldots}{\forall^m_m} \vdash f^n_m \vdash \underset{ij\ldots}{\exists^m_m}.$$

Indeed, let R be an m-ary relation; if

$$\mathrm{R}(x_1,\ldots, x',\ldots, x'',\ldots, x_m) = +$$

for all values x', x'',... substituted for x_i, x_j,..., then, in particular,

$$\mathrm{R}(x_1,\ldots, x_{f(i)},\ldots, x_{f(j)},\ldots, x_m) = +.$$

And if the last equality is satisfied, there exist a value x', a value x'',... such that

$$\mathrm{R}(x_1,\ldots, x',\ldots, x'',\ldots, x_m) = +.$$

3.8.4. *Let i, j,... be integers, f and f' two mappings defined on* $\{1,\ldots, m\}$. *Assume that* $f(u) = f'(u)$ *for all* $u \leqslant m$, $\neq i, j,\ldots$, *and i, j,... are inactive in the relation* R. *Then* $f^n_m\mathrm{R} - \vdash f'^{n'}_m\mathrm{R}$ *for all* $n \geqslant f(1),\ldots, f(m)$ *and* $n' \geqslant f'(1),\ldots, f'(m)$.

Indeed, the first member takes the value

$$\mathrm{R}(x_{f(1)},\ldots, x_{f(i)},\ldots, x_{f(j)},\ldots, x_{f(m)})$$

and the second takes the value

$$\mathrm{R}(x_{f(1)},\ldots, x_{f'(i)},\ldots, x_{f'(j)},\ldots, x_{f(m)}).$$

3.8.5. (1) *Let f be a mapping of* $\{1,\ldots, m\}$ *into* $\{1,\ldots, n\}$ *and let* $i \leqslant m$. *Denote by f' the mapping of* $\{1,\ldots, m\}$ *into* $\{1,\ldots, n+1\}$ *such that* $f'(u) = f(u)$ *for all* $u \neq i$, *and* $f'(i) = n+1$; *then*

$$f^n_m \underset{i}{\forall^m_m} = \underset{n+1}{\forall^n_{n+1}} f'^{n+1}_m.$$

▷ Since f and f' differ only for i and the index i is inactive in $\underset{i}{\forall^m_m}\mathrm{R}$ for all R, it follows from 3.8.4 that $f^n_m \underset{i}{\forall^m_m} - \vdash f'^{n+1}_m \underset{i}{\forall^m_m}$. Furthermore, f' takes the value $f'(i) = n+1$ only for i; it therefore follows from 3.8.1 that

$$f'^{n+1}_m \underset{i}{\forall^m_m} = \underset{n+1}{\forall^{n+1}_{n+1}} f'^{n+1}_m - \vdash \underset{n+1}{\forall^n_{n+1}} f'^{n+1}_m.$$

Thus, we finally get $f^n_m \underset{i}{\forall^m_m} \dashv\vdash \underset{n+1}{\forall^n_{n+1}} f'^{n+1}_m$, and these equideducible operators have the same arity; they are thus identical. ◁

(2) *Let f be the same mapping, and let f'' denote the mapping of $\{1, \ldots, m+1\}$ into $\{1, \ldots, n+1\}$ such that $f''(u) = f(u)$ for $u = 1, \ldots, m$ and $f''(m+1) = n+1$; then*

$$f^n_m \underset{m+1}{\forall^m_{m+1}} = \underset{n+1}{\forall^n_{n+1}} f''^{n+1}_{m+1}.$$

▷ Let d^{m+1}_m denote the dilator which associates with each m-ary relation R the relation

$$\mathrm{R}'(x_1, \ldots, x_m, x_{m+1}) = \mathrm{R}(x_1, \ldots, x_m) \quad \text{(see 3.4.5)}.$$

Then $f''^{n+1}_{m+1} d^{m+1}_m \dashv\vdash f^n_m$; indeed, each of these operators takes the value + for R and $(x_1, \ldots, x_n)$, whenever $\mathrm{R}(x_{f(1)}, \ldots, x_{f(m)}) = +$. We thus have

$$f^n_m \underset{m+1}{\forall^m_{m+1}} \dashv\vdash f''^{n+1}_{m+1} d^{m+1}_m \underset{m+1}{\forall^m_{m+1}};$$

but $d^{m+1}_m \underset{m+1}{\forall^m_{m+1}} = \underset{m+1}{\forall^{m+1}_{m+1}}$, so that the first operator considered is equideducible from

$$f''^{n+1}_{m+1} \underset{m+1}{\forall^{m+1}_{m+1}} = \underset{n+1}{\forall^{n+1}_{n+1}} f''^{n+1}_{m+1},$$

thus equideducible from $\underset{n+1}{\forall^n_{n+1}} f''^{n+1}_{m+1}$, and of the same arity. ◁

3.8.6. *Let $\mathscr{U}$ be a composite (m, m)-ary quantifier in which the indices decrease in the order of composition* (hence increase in the usual notation). *Let f be a function defined on the integers $i = 1$ to m, such that:*

$f(i) \leqslant i$ if i is the index of a quantifier $\exists$;

$f(i) = i$ if i is the index of a quantifier $\forall$ or if i is not an index of $\mathscr{U}$. Then:

(1) *$\mathscr{U} f^m_m \vdash \mathscr{U}$.*

(2) *If $n = \max(f(1), \ldots, f(m))$ and $\mathscr{U}'$ is obtained from $\mathscr{U}$ by replacing the arity m by n and suppressing the quantifiers of indices $n+1$ to m, then $\mathscr{U}' f^n_m \vdash \mathscr{U}$.*

▷(1) First consider the special case in which, for some index i of a quantifier $\exists$, we have $f(i) < i$ and $f(u) = u$ for all $u \neq i$. Decompose $\mathscr{U}$ as

$$\left(\underset{1}{\mathrm{U}}\,\underset{2}{\mathrm{U}} \ldots \underset{i-1}{\mathrm{U}}\right)^m_m \underset{i}{\exists^m_m} \left(\underset{i+1}{\mathrm{U}} \ldots \underset{m}{\mathrm{U}}\right)^m_m,$$

in which each letter U stands for a simple quantifier $\forall$ or $\exists$. By 3.8.3, we have $f^m_m \vdash \underset{i}{\exists}{}^m_m$, and so

$$f^m_m \left(\underset{i+1}{\mathrm{U}} \ldots \underset{m}{\mathrm{U}} \right)^m_m \vdash \left(\underset{i}{\exists} \underset{i+1}{\mathrm{U}} \ldots \underset{m}{\mathrm{U}} \right)^m_m.$$

The function f takes each of the values from $i+1$ to m only once: indeed, these are values $>i$, and therefore also $>f(i)$.

By 3.8.1, in view of the fact that

$$f(i+1) = i+1, \ldots, f(m) = m,$$

we have

$$f^m_m \left(\underset{i+1}{\mathrm{U}} \ldots \underset{m}{\mathrm{U}} \right)^m_m = \left(\underset{i+1}{\mathrm{U}} \ldots \underset{m}{\mathrm{U}} \right)^m_m f^m_m,$$

and hence

$$\left(\underset{i+1}{\mathrm{U}} \ldots \underset{m}{\mathrm{U}} \right)^m_m f^m_m \vdash \left(\underset{i}{\exists} \underset{i+1}{\mathrm{U}} \ldots \underset{m}{\mathrm{U}} \right)^m_m,$$

which gives

$$\left(\underset{i}{\exists} \underset{i+1}{\mathrm{U}} \ldots \underset{m}{\mathrm{U}} \right)^m_m f^m_m \vdash \left(\underset{i}{\exists} \underset{i+1}{\mathrm{U}} \ldots \underset{m}{\mathrm{U}} \right)^m_m,$$

and finally

$$\left(\underset{1}{\mathrm{U}} \ldots \underset{i-1}{\mathrm{U}} \underset{i}{\exists} \underset{i+1}{\mathrm{U}} \ldots \underset{m}{\mathrm{U}} \right)^m_m f^m_m \vdash \left(\underset{1}{\mathrm{U}} \ldots \underset{i-1}{\mathrm{U}} \underset{i}{\exists} \underset{i+1}{\mathrm{U}} \ldots \underset{m}{\mathrm{U}} \right)^m_m,$$

which is (1) for the special case under consideration.

In the general case, decompose f as a product of functions $f_i f_j \ldots = f$, in which each f_i is the identity for every integer other than i. We get $\mathcal{U} f_i \vdash \mathcal{U}$, so that $\mathcal{U} f_i f_j \vdash \mathcal{U} f_j \vdash \mathcal{U}$, and so (1) finally follows.

(2) Since $f(i) \leqslant i$, we have $n = \max(f(1), \ldots, f(m)) \leqslant m$ and $f^n_m \dashv\vdash f^m_m$. With each quantifier $\underset{i}{\mathrm{U}}{}^m_m$ we associate $\underset{i}{\mathrm{U}}{}^n_n$, which differs from it only in the change of arity from m to n; then

$$\left(\underset{1}{\mathrm{U}} \ldots \underset{n}{\mathrm{U}} \right)^n_n f^n_m \dashv\vdash \left(\underset{1}{\mathrm{U}} \ldots \underset{m}{\mathrm{U}} \right)^m_m f^m_m,$$

in other words, $\mathcal{U}' f^n_m \dashv\vdash \mathcal{U} f^n_m \vdash \mathcal{U}$, which is (2). ◁

Example. Take $\mathcal{U} = (\underset{1\,2\,3\,4}{\forall\exists\forall\exists})^4_4$ and $f(1)=1, f(2)=1, f(3)=3$ and $f(4)=2$.

We get the deductions

$$\begin{pmatrix} \forall\exists\forall \\ 1\,2\,3 \end{pmatrix}^3_3 f^3_4 \dashv\vdash \begin{pmatrix} \forall\exists\forall\exists \\ 1\,2\,3\,4 \end{pmatrix}^4_4 f^4_4 \vdash \begin{pmatrix} \forall\exists\forall\exists \\ 1\,2\,3\,4 \end{pmatrix}^4_4 .$$

In other words, given a quaternary relation R, if for any x_1 there exists x_2 such that we have $R(x_1, x_1, x_3, x_2) = +$ for any x_3, then for any x_1 there exists x_2 such that for any x_3 there exists x_4 satisfying $R(x_1, x_2, x_3, x_4) = +$.

EXERCISES

1

Let R and S be two n-ary relations. Define $R \leqslant S$ if there exists a restriction of S isomorphic to R; set $R \gtrless S$ if $R \leqslant S$ and $S \leqslant R$; finally, set $R < S$ if $R \leqslant S$ but not $S \leqslant R$.

(1) Show that $\leqslant$ is a pre-ordering (reflexive and transitive relation) on the n-ary relations, and hence that $\gtrless$ is an equivalence relation.

(2) Recall that a *chain* (or total ordering) is a binary relation that is reflexive, transitive and such that any two elements of the base are comparable. Let Q denote the chain of rationals (of arbitrary sign). Show that every denumerable chain is $\leqslant Q$. Show that a denumerable chain is isomorphic to Q if and only if it is dense (i.e., between any two elements there is a third element) and has neither minimum nor maximum. Give an example of two chains which are $\gtrless$ Q but not isomorphic; give an example of two nonisomorphic chains $R \gtrless S < Q$.

(3) Show that every denumerable chain is either $\geqslant$ the chain N of positive integers or $\geqslant$ the chain N′ of negative integers.

(4) Starting from the chain N and replacing each positive integer in N by a chain which is either finite or isomorphic to the chain of all integers (positive, zero, and negative), show that the set of pairwise nonisomorphic and equivalent (in the sense $\gtrless$) chains has the cardinality of the continuum.

(5) Starting from the chain N, replace each integer by a chain $< Q$; show that the resulting chain is $< Q$. In particular, if $R < Q$ and $S < Q$, then $R + S < Q$.

(6) If R is a denumerable chain which is $\geqslant$ every denumerable well-ordering, then $R \gtrless Q$. Deduce that for every denumerable chain $R < Q$ there exists R′ such that $R < R' < Q$.

(7) Associate with each denumerable ordinal α a denumerable chain $R(\alpha)$ such that

(i) if $\alpha < \beta$, then $R(\alpha) < R(\beta)$;

(ii) for all α and β, there exists γ such that $R(\gamma) \geqslant R(\alpha) + R(\beta)$.

Then any denumerable chain $X \geqslant$ all $R(\alpha)$ is $\gtrless Q$. Show that condition (ii) alone is not sufficient for the conclusion to hold, even if we require that each $R(\alpha)$ be infinite.

Results of [LAV, 1971]. Every strictly decreasing sequence of denumerable chains is finite (for chains of the cardinality of the continuum, an infinite sequence was constructed by [DUS-MIL, 1940] on the basis of the chain of real numbers).

Every set of pairwise incomparable (with respect to $\leqslant$) denumerable chains is finite.

Condition (i) is sufficient to ensure that any denumerable chain which is $\geqslant$ all $R(\alpha)$ be $\gtrless$ Q. Given a chain $R < Q$, the equivalence classes induced by $\gtrless$ on the chains $\leqslant R$ constitute an at most denumerable set.

2

(1) Retaining the notation $\leqslant$ of Exercise 1 and using the classical proof of the Cantor-Bernstein theorem, prove the following proposition:

If $R \leqslant S$ and $S \leqslant R$, then there exist a partition of the base E of R into E′ and E″ ($E' \cup E'' = E$, $E' \cap E'' = \emptyset$) and a partition of the base F of S into F′ and F″, such that $R \mid E'$ is isomorphic to $S \mid F'$ and $R \mid E''$ is isomorphic to $S \mid F''$.

Show that the converse is false; that is to say, give an example of a relation R having base E and another relation S having base F, such that R is not comparable with S (i.e., is neither $\leqslant S$ nor $\geqslant S$) but $R \mid E'$ is isomorphic to $S \mid F'$ and $R \mid E''$ is isomorphic to $S \mid F''$; give another example, with $R \leqslant S$ but not $S \leqslant R$.

(2) Let R_p be the binary relation defined on $\{1, \ldots, p\}$ (where p is a positive integer) which is equal to + for the pairs (1, 2), (2, 3), ..., $(p-1, p)$ and $(p, 1)$. Show that for $p \neq q$ the

relations R_p and R_q are not comparable. Hence deduce that the set of pairwise incomparable denumerable binary relations has the cardinality of the continuum. Prove the same for the n-ary relations ($n \geqslant 3$).

(3) Show that for each n there exist denumerable n-ary relations which are $\geqslant$ every denumerable n-ary relation. Except for $n = 1$, these are not all isomorphic.

(4) Given two n-ary relations R and S with finite bases such that R is not $\geqslant$S, show that there exists an extension of R to the set obtained by adding a single element to the base, and hence an extension of R of arbitrarily large cardinality (finite or infinite), which is still not $\geqslant$S.

(5) Given a finite relation R and a finite set of finite relations $U_1, \ldots, U_h$, with R not $\geqslant U_i$ for any $i = 1, \ldots, h$, there does not always exist a strict extension of R which satisfies the same conditions.

Problem of the compatible extension. Given finite relations $U_1, \ldots, U_h$, does there exist an integer p such that every binary relation of cardinality $\geqslant p$ which is not $\geqslant U_i$ for any $i = 1, \ldots, h$ has a proper extension (hence an extension of arbitrarily large cardinality) which satisfies the same conditions? J. Malitz has shown that in the general case the answer is negative (see 4.8.6). For binary relations, negative answer by G. Lopez.

3

Consider a finite or denumerable set of relations A of the same arity, over finite bases. Let the set be closed under embeddability $\leqslant$ (see Exercise 1): every relation isomorphic to a restriction of an A is an A. Let us say that a set of relations B from the set of relations A is an *embedding set* if (i) the relations B are pairwise incomparable (with respect to $\leqslant$); (ii) for every A, there exists a B such that B$\geqslant$A.

(1) Show that if the set of relations A is finite, there exists an embedding set. Give an example of an infinite set of relations for which an embedding set exists, and another example for which no embedding set exists. Show that the embedding set, if it exists, is unique (up to isomorphism of relations), and that it is the set of maximal relations under the embeddability ordering.

(2) Starting from the set of relations A (which is closed under embeddability), eliminate the maximal relations, and iterate the elimination procedure infinitely many times. The residual set is closed under embeddability; denote the relations in this set by C (they are of course A's). Show that the relations C constitute the largest set of relations A which is closed under embeddability and contains no maximal relations.

(3) Consider a set of relations C with finite bases, closed under embeddability, containing no maximal relations. Give an example of a set of this type for which there exist(s) one or finitely many relations R over denumerable bases, such that the restrictions of R's to finite subsets of their base(s) yield (up to isomorphism) all relations C. Show that there exists a minimal set of relations R, in the sense that when any of the R's is eliminated the finite restrictions do not constitute all the C's. Moreover, the number of relations R is the same for any minimal set with finite restrictions C, and there is a bijective correspondence between any two minimal sets such that two corresponding R's have the same finite restrictions.

(4) Find a set of finite relations C closed under embeddability and containing no maximal relations, for which the above construction requires infinitely many R's; note that we can limit ourselves to a denumerably infinite set. Show that, for the following example, there exists no maximal set of relations R in the sense of (3) above: For each integer n, let the base consist of all integers $0, \ldots, n$, and consider all relations C_n satisfying $C_n(0, 1) = C_n(1, 0) = +$, then $C_n(x, y) = +$ for $y = x+1$ and $-$ for $y \neq x$ and $\neq x+1$ (except for $x = 1$

and $y=0$); finally, define $C_n(x, x)$ arbitrarily for all $x=0, \ldots, n$. Now add all restrictions of the relations C_n. Give another example, for which infinitely many R's are needed, but there exists a minimal set of R's.

(5) Show that for a given set of relations C there cannot be both a finite minimal set and an infinite minimal set. There is a bijective correspondence between any two infinite minimal sets such that two corresponding R's have the same finite restrictions. In a minimal set of R's, no subset of several relations R can be replaced by a single relation yielding the same finite restrictions.

More precisely, R (or a relation having the same age) belongs to a minimal set if and only if the finite restrictions of R (in the set of relations C) constitute a maximal finite-age which is not contained in the union of other maximal finite-ages, and moreover each C belongs to a finite-age satisfying this condition.

(6) If each set of C's which is incomparable under embeddability is finite, there exists a finite set of R's (parts (5) and (6) were communicated to us by M. Pouzet).

4

(1) Divide the (non-ordered) pairs of natural numbers into two classes. Show that there exists an infinite set of integers a_i such that all pairs $\{a_i, a_j\}$ $(i, j=0, 1, 2, \ldots)$ belong to the same class. (First construct a sequence $a_i (i=0, 1, 2, \ldots)$ such that, for each i, all pairs $\{a_i, a_{i+1}\}, \{a_i, a_{i+2}\}, \ldots$ belong to the same class (depending on i).)

Generalize the preceding result to the case of sets of p natural numbers, still assuming these sets to be divided into two classes. Then assume them to be divided into a finite number of classes [RAM, 1926, 1929].

(2) Show that, for given integers a, p, k with $a \geqslant p$, there exists $a' \geqslant a$ such that, for every set E of cardinality at least a' whose subsets of cardinality p are divided into k classes, there exists a subset F of E, of cardinality a, such that the subsets of F of cardinality p all belong to the same class (*loc. cit.*). Show that for $p=1$ we can take $a'=k(a-1)+1$; for $k=1$ we can take $a'=a$; and for $a=p$ we can take $a'=p$. Show that for $p=k=2$ and $a=3$ we can take $a'=6$ but not 5; for $p=2$, $k=3$, $a=3$ we can take $a'=17$.

(3) Utilizing the axiom of choice and well-ordering the reals, divide the pairs of reals into two classes in such a way that if A is a set in which every pair belongs to the same class, then A is finite or denumerable. Hence deduce that the assertion in (1) fails to hold if the set of integers is replaced by a set with the cardinality of the continuum [SIE, 1933].

(4) Derive the following propositions from (1):

From every infinite sequence of real numbers one can extract an increasing, decreasing or constant infinite sequence.

Given two well-ordered sets of real numbers, say u_i and v_j, the set of sums u_i+v_j is well-ordered.

(5) Divide the strictly positive integers into a finite number of classes; then at least one of these classes will include three distinct integers a, b, c such that $c=a+b$ [SCHU, 1916]. (*Hint.* With each class A of integers, associate the class A^* of pairs $\{u, v\}$ of integers such that $|u-v| \in A$; then apply the theorems of (1) and (2) to A^*.)

(6) Show that every denumerable relation has a denumerable restriction, all of whose finite restrictions having the same number of elements are isomorphic.

5 (see [ERD-RAD, 1950])

(1) Assume the (non-ordered) pairs of natural numbers to be divided into a finite or infinite number of equivalence classes. Show that there exists an infinite set F of integers such that one of the following four assertions holds:

The pairs contained in F are all equivalent.
They are all from distinct classes.
They are equivalent whenever they have the same first element.
They are equivalent whenever they have the same second element.
Hint. Consider the relation defined by

$$R(x_1, x_2, x_3, x_4) = +$$

if and only if

$$x_1 < x_2 \quad \text{and} \quad x_3 < x_4 \quad \text{and} \quad \{x_1, x_2\} \quad \text{is equivalent to} \quad \{x_3, x_4\}$$

and apply Ramsey's theorem (see Exercise 4).
(2) Generalize to sets of p elements (p a positive integer).

6

Given a finite number of relations $R_1, \ldots, R_p$ with common base E, denote by $\bigwedge(R_1, \ldots, R_p)$ the relation whose arity is the maximum of those of $R_1, \ldots, R_p$, defined as + for those sequences which give the value + to all of $R_1, \ldots, R_p$. Given a set $\mathscr{A}$ of relations with base E and a relation S, define $\mathscr{A} \vdash S$ if there exist in $\mathscr{A}$ finitely many relations $R_1, \ldots, R_p$ such that $\bigwedge(R_1, \ldots, R_p) \vdash S$. Define $\mathscr{A} \vdash \mathscr{B}$ whenever $\mathscr{A} \vdash S$ for all $S \in \mathscr{B}$.
(1) Show that $\vdash$ is a pre-ordering of the sets of relations with base E.
(2) Let $\overline{\mathscr{A}}$ denote the set of S such that $\mathscr{A} \vdash S$. Show that the operation $\overline{}$ has the following properties: (a) $\overline{\mathscr{A}} \supseteq \mathscr{A}$; (b) $\overline{\overline{\mathscr{A}}} = \overline{\mathscr{A}}$ (idempotence); (c) if $\mathscr{A} \subseteq \mathscr{B}$ then $\overline{\mathscr{A}} \subseteq \overline{\mathscr{B}}$ (monotonicity); (d) $\overline{\mathscr{A}}$ is the union of the sets $\overline{\mathscr{F}}$ for all finite subsets $\mathscr{F} \subseteq \mathscr{A}$.
(3) What is the set $\overline{\emptyset}$ generated by the empty set $\emptyset$? Deduce from (2) that

$$\overline{\mathscr{A} \cap \mathscr{B}} \subseteq \overline{\mathscr{A}} \cap \overline{\mathscr{B}} \quad \text{and} \quad \overline{\mathscr{A} \cup \mathscr{B}} \supseteq \overline{\mathscr{A}} \cup \overline{\mathscr{B}}.$$

Give examples in which the first and second members of these inclusions are distinct.
In general,

$$\overline{\bigcap_i \mathscr{A}_i} \subseteq \bigcap_i \overline{\mathscr{A}_i} \quad \text{and} \quad \overline{\bigcup_i \mathscr{A}_i} \supseteq \bigcup_i \overline{\mathscr{A}_i}.$$

If every union of a finite set of $\mathscr{A}_i$ is contained in an $\mathscr{A}_i$, then

$$\bigcup_i \overline{\mathscr{A}_i} = \overline{\bigcup_i \mathscr{A}_i}.$$

7

(1) Let M and M′ be two multirelations with bases E and E′; assume that every finite restriction of M′ is isomorphic to a restriction of M. Show that for every N with base E there exists N′ with base E′ such that every finite restriction of M′N′ is isomorphic to a restriction of MN (the concatenations MN and M′N′ are defined in 3.1).
Hint. With each finite subset F of E′, associate the set of isomorphisms u of M′/F onto a restriction of M; then, for each u consider the image $N'_{F,u}$ of $N/u(F)$ under the inverse u^{-1}. Now, with F remaining fixed and u varying, consider the set U_F of extensions to E′ of the multirelations $N'_{F,u}$. Note that the supersets of the U_F form a filter over the multirelations with base E′, and employ the ultrafilter axiom.
(2) Let M have base E and let M′ be an extension of M to a set $E' \supseteq E$. We say that M′ is a 1-*extension* of M if, for every finite subset F of E′, there exists an isomorphism of the restriction $M' \mid F$ onto a restriction of M which reduces to the identity on the intersection

F∩E. Again using the ultrafilter axiom, show that for all N with base E there exists an N′ with base E′ such that M′N′ is a 1-extension of MN.

(3) If R, S are two infinite chains, there exist an S′ isomorphic to S and a common 1-extention to R and S′.

8

Given an n-ary relation R, a *bound* of R is any relation A with finite base such that A is not $\leqslant$R but every strict restriction of A is $\leqslant$R (recall that $\leqslant$ means "isomorphic to a restriction of"). Note that bounds are pairwise incomparable with respect to the ordering $\leqslant$.

(1) Show that if X is finite, then X$\leqslant$R if and only if X is not $\geqslant$ any bound of R.

(2) Show that a chain (total ordering) has only a finite number of bounds (up to isomorphism); represent each of these by a table of + and − signs. On the other hand, show that the successor relation, defined as + for pairs (x, y) of natural numbers such that $y=x+1$, has denumerably many bounds.

(3) Show that two relations have the same finite restrictions (up to isomorphism) if and only if they have the same bounds.

(4) If R is a relation of finite cardinality p, with a bound A, show that there exists an extension of R of cardinality $p+1$ for which A is also a bound.

(5) Given a finite number of finite relations $A_1, \ldots, A_h$, if there exist relations of arbitrarily large cardinality, each of which has bounds $A_1, \ldots, A_h$ (and possibly others), show that there exists a denumerable relation which has these bounds (and possibly others).

(6) Given a finite number of finite relations $A_1, \ldots, A_h$, suppose that for every integer p there exists a relation of cardinality $\geqslant p$ whose bounds are either $A_1, \ldots, A_h$ or of cardinality $\geqslant p$. Show that there exists a denumerable relation H whose bounds are exactly $A_1, \ldots, A_h$. For any such relation H and any finite relation R, we have either R$\geqslant A_1$, or ..., or R$\geqslant A_h$, or R$\leqslant$H (part (6) was suggested by J.-F. Pabion).

9

Let E be a set, m an integer, and $\mathscr{A}$ a set of permutations of E. A permutation f of E is said to be *m-adherent* to $\mathscr{A}$ if, for any $a_1, \ldots, a_m$ of E, there exists a permutation in $\mathscr{A}$ which maps a_1 onto $f(a_1), \ldots, a_m$ onto $f(a_m)$.

(1) If R is an m-ary relation, or, in general, a (finite or infinite) sequence of m-ary relations with base E, the group of automorphisms of R is closed under m-adherence: every permutation which is m-adherent to the group is an element of the group. Prove the converse.

(2) Every set of permutations which is closed under m-adherence is also closed under m'-adherence for every $m' \geqslant m$. Considering a set E of cardinality $m+2$ and an odd permutation f of E, show that the set of permutations ft, where t is a transposition of any two elements of E, generates a group which is closed under $(m+1)$-adherence but not under m-adherence.

(3) Let $m \geqslant 2$; let E be the set of rational points of dimension $m+1$, and $\mathscr{A}$ the group of linear permutations of E with positive determinant. Show that $\mathscr{A}$ is closed under $(m+1)$-adherence, but there is a linear permutation with negative determinant which is m-adherent to $\mathscr{A}$ (parts (2) and (3) were communicated by M. Pouzet).

(4) A group of permutations of E is said to be *m-transitive* if every permutation of E is m-adherent to the group. Note that the symmetric group on E is m-transitive for any m. If E is finite with cardinality $\geqslant m+2$, the alternating group on E (group of even permutations) is m-transitive. It is not known if any m-transitive group ($m \geqslant 6$) is necessarily symmetric or alternating.

CHAPTER 4

LOCAL ISOMORPHISM; FREE OPERATOR AND FREE FORMULA

4.1. Local isomorphism; local automorphism

Given two multirelations M over E and M′ over E′, of the same arity, we define a *local isomorphism* of M *onto* M′ as any isomorphism of a restriction of M onto a restriction of M′.

Examples. A bijective mapping f of a subset of the base of an ordering S onto a subset of the base of an ordering S′, with $f(x)$ preceding $f(y)$ in S′ if and only if x precedes y in S.

A bijective mapping f of a subset of the base of an ordered group (R, S) onto a subset of the base of another ordered group (R′, S′), with $f(x)\,f(y)=f(z)$ in the group R′ if and only if $xy=z$ in the group R, and also the preceding condition for the orderings S and S′.

4.1.1. *A mapping f is a local isomorphism of* $M=(R_1,\ldots,R_h)$ *onto* $M'=(R'_1,\ldots,R'_h)$ *if and only if it is a local isomorphism of* R_k *onto* R'_k *for each* $k=1,\ldots,h$.

Indeed, let F be the set over which f is defined; then

$$M \mid F=(R_1 \mid F,\ldots, R_h \mid F)$$

implies

$$f(M \mid F)=(f(R_1 \mid F),\ldots,f(R_h \mid F))$$

and

$$M' \mid f(F)=(R'_1 \mid f(F),\ldots, R'_h \mid f(F)).$$

It thus follows that $f(M \mid F)=M' \mid f(F)$ if and only if

$$f(R_i \mid F)=R'_i \mid f(F) \quad \text{for} \quad i=1,\ldots,h.$$

It follows from our previous conventions that:

$f_\emptyset$ is a local isomorphism of every m-ary relation (m positive) onto every other m-ary relation;

$f_\emptyset$ is a local isomorphism of 0-ary relation (E, +) onto (E′, +) and of (E, −) onto (E′, −) for every E and E′; *but it is not a local isomorphism of* (E, +) *onto* (E′, −), nor of (E, −) onto (E′, +);

$f_\emptyset$ is a local isomorphism of M onto a multirelation M′ of the same arity if and only if, to each 0-ary relation R_k in M, there corresponds a 0-ary relation R'_k of the same rank in M′, having the same value (+ or −).

4.1.2. *If f, defined on* F, *is a local isomorphism of* M *onto* M′, *then the restriction of f to an arbitrary subset of* F *is a local isomorphism of* M *onto* M′.

An isomorphism f of M onto M′, and consequently the restriction of f to any subset of the base of M, is a local isomorphism of M onto M′.

In particular, for every subset F of the base, the identity on F is a local isomorphism on M onto itself.

On the other hand, a local isomorphism of M onto M′ cannot in general be extended to an isomorphism of M onto M′ (even if the bases of M and M′ are equipollent, and even if M and M′ are isomorphic or even identical). For example, if R is the chain (total ordering) of nonnegative integers and R′ the chain of nonpositive integers, then the mapping f which takes 0, 1, 2, ..., p onto

$$f(0)=-p, \qquad f(1)=-p+1, \qquad \ldots, \qquad f(p)=0,$$

is a local isomorphism of R onto R′. It cannot be extended to an isomorphism $\bar{f}$ of R onto R′, because R and R′ are not isomorphic. Moreover, otherwise the integer x mapped onto $\bar{f}(x)=-p-1$ would have to be a predecessor of 0 in R.

If R is the chain of nonnegative integers and R′ = R, then the mapping f which takes 0 onto $f(0)=1$ cannot be extended to an isomorphism $\bar{f}$ of R onto itself, since otherwise the integer x mapped onto $\bar{f}(x)=0$ would have to be a predecessor of 0 in R.

4.1.3. *If f is a local isomorphism of* M *onto* M′, *then* f^{-1} *is a local isomorphism of* M′ *onto* M.

If in addition g is a local isomorphism of M′ *onto* M″, *then the composite mapping gf is a local isomorphism of* M *onto* M″.

4.1.4. *Let* R *and* R′ *be two m-ary relations; then a bijective mapping f defined on* F *is a local isomorphism of* R *onto* R′ *if and only if, for every subset* G *containing at most m elements of* F, *the restriction of f to* G *is a local isomorphism of* R *onto* R′.

Let M and M′ have the same arity $(m_1, \ldots, m_h)$, and let m be the maximum of $m_1, \ldots, m_h$.

A sufficient condition for f to be a local isomorphism of M *onto* M′ *is that, for every* G *containing at most m elements, the restriction of f to* G *be a local isomorphism of* M *onto* M′.

Proof of the first assertion. Let $x_1, \ldots, x_m$ be an m-tuple over F. The set $G = \{x_1, \ldots, x_m\}$ contains at most m elements; by assumption, we have

$$R'(f(x_1), \ldots, f(x_m)) = R(x_1, \ldots, x_m).$$

We get another proof of the assertions of 3.1.1 by letting f be the identity on the base E of R (or of M).

4.1.5. We define a *local automorphism* of M to be any local isomorphism of M onto itself. An automorphism f of M (isomorphism of M onto itself) and, consequently, the restriction of f to any subset of the base, is a local automorphism of M. On the other hand, a local automorphism cannot always be extended to an automorphism of M (see the example of 4.1.2 with R′ = R).

4.2. Free Interpretability

Let M and N be two multirelations over the same base E. We shall say that N is *freely interpretable* in M if *every local automorphism of* M *is a local automorphism of* N.

Note that it is sufficient for every local automorphism of M *defined on at most n elements* (n = maximum arity of N) to be a local automorphism of N; see 4.1.4.

Examples. Let $R(x, y)$ be an ordering; the strict ordering relation: $S(x, y) = +$ if x precedes y and is distinct from it, is freely interpretable in R, and R is freely interpretable in S. The relation: $T(x, y, z) = +$ if z is between x and y in the ordering R, is freely interpretable in R. In fact, every bijective mapping f of a subset of the base onto another which preserves the ordering of the elements also preserves the strict ordering and the "between" relation. However, R is not freely interpretable in T: consider the ordering of the integers 1, 2, 3 and let f be the permutation which inverts this order, mapping 1, 2, 3 onto 3, 2, 1.

If R is a chain (total ordering), the cycle associated with R will be the relation: $S(x, y, z) = +$ whenever $x \leqslant y \leqslant z$ or $y \leqslant z \leqslant x$ or $z \leqslant x \leqslant y$ in R. We see that S is freely interpretable in R but that R is not freely interpretable in S: the transposition exchanging two different elements x, y, is a local automorphism of S, not of R.

Free interpretability is a pre-ordering (reflexive and transitive), but not an ordering. Indeed, two different relations can be freely interpretable in each other. *Example*: R and the relation which has the same base as R and always takes the opposite value (this relation has been denoted by $\neg R$). As we have seen, another example is an ordering and the corresponding strict ordering.

A multirelation which is freely interpretable in M is *a fortiori* freely interpretable in any multirelation obtained by adjoining arbitrary relations (having the same base as M) to M.

N *is freely interpretable in* M *if and only if each relation of* N *is freely interpretable in* M. In particular, every sequence of relations extracted from M (with possible repetitions) is freely interpretable in M.

The m-ary relation on E identically equal to $+$, the relation identically equal to $-$, and the identity relation on E, i.e., $R(x, y) = +$ when $x = y$ and $-$ when $x \neq y$ (for $x, y \in E$), are all freely interpretable in M, for every M over E.

4.2.1. *For every positive integer n, there exists a denumerable n-ary relation which is not freely interpretable in any multirelation whose maximum arity is smaller than n.*

▷ Let the base be the set of positive integers. For every positive integer p and n-ary relation A over the set of integers $\{1, 2, \ldots, p\}$, let $u(p, A)$ be an integer such that the sets $F(p, A) = \{u(p, A)+1, \ldots, u(p, A)+p\}$ are all disjoint for distinct ordered pairs (p, A). This is possible, since there are finitely many relations A for each p – exactly 2^{p^n}. For each ordered pair (p, A), consider the image of A under the bijective mapping which takes each $x = 1, 2, \ldots, p$ onto $u(p, A)+x$. The base of this new relation is $F(p, A)$. Let R be an extension of all these relations; we now show that R satisfies the assertion. Suppose that there exist a positive integer h, integers m_i $(i = 1, 2, \ldots, h)$ equal to at most $n-1$, and an $(m_1, \ldots, m_h)$-ary multirelation M such that R is freely interpretable in M. The number of multirelations over $\{1,\ldots, p\}$ which have this arity is 2 to the power of $p^{m_1}+\cdots+p^{m_h}$, which is $\leqslant 2^{hp^{n-1}}$ and hence $<2^{p^n}$, provided $p>h$. Consider such a value of p, and the sets $F(p, A)$ for all relations A which correspond to p. For two distinct relations A, say A and B, the bijective mapping of $F(p, A)$ onto $F(p, B)$ which preserves the order of the integers can be factored into the product of the mapping $x-u(p, A)$, which maps the restriction $R \mid F(p, A)$ onto A, and the mapping $x+u(p, B)$, which maps B onto the restriction $R \mid F(p, B)$. Since A and B are distinct, the composite mapping is not a local automorphism of R. On the other hand, since the number of $(m_1,\ldots, m_h)$-ary multirelations over $\{1,\ldots, p\}$ is smaller than the number of relations A over the same set, there exists a pair of distinct relations A and B such that the image under $x-u(p, A)$ of the restriction $M \mid F(p, A)$ and the image under $x+u(p, B)$ of $M \mid F(p, B)$ are identical. Then the composite mapping $x-u(p, A)+u(p, B)$ is a local automorphism of M but not of R (communicated by J. F. Pabion). ◁

4.2.2. Let M and N be multirelations over a base E, of arities μ and ν, respectively, and M′ and N′ multirelations over E′ of the same respective arities. Let n denote the maximum of the arity ν.

If every restriction of M′N′ *to at most n elements is isomorphic to a restriction of* MN *and* N *is freely interpretable in* M, *then* N′ *is freely interpretable in* M′ (the concatenations MN and M′N′ are defined in 3.1; communicated by J. L. Paillet).

▷ Let f' be a local automorphism of M′, defined on a set F′ with at most n elements. By hypothesis, there exists an isomorphism g of the restriction $(M'N') \mid F'$ onto a restriction of MN, and an isomorphism h of $(M'N') \mid f'(F')$ onto a restriction of MN. Let f be $hf'g^{-1}$; we see that f is a local automorphism of M, and hence of N, which is freely interpretable in M. Hence f' is a local automorphism of N′. Thus every local automorphism of M′ is a local automorphism of N′ (see 4.2). ◁

For example, we have seen that if R is a chain and S the associated ternary cycle ($S(x, y, z)=+$ whenever $x\leqslant y\leqslant z$ or $y\leqslant z\leqslant x$ or $z\leqslant x\leqslant y$ for R), then this cycle is freely interpretable in R. Consider another chain R′ and a ternary relation S′ such that every restriction of R′S′ to at most 3 elements is isomorphic to a restriction of RS; then S′ is the cycle associated with R′, which we know is freely interpretable in R′.

4.3. Free operator

A (μ, n)-ary operator $\mathscr{P}$ is said to be *free* if for any μ-ary multirelations M and M′ *every local isomorphism of* M *onto* M′ *is a local isomorphism of* $\mathscr{P}$(M) onto $\mathscr{P}$(M′).

An equivalent definition is: for every subset F of the base of M, we have $\mathscr{P}(\mathrm{M} \mid \mathrm{F}) = \mathscr{P}(\mathrm{M}) \mid \mathrm{F}$, and for every bijective mapping f defined on the base of M we have $\mathscr{P}(f(\mathrm{M})) = f(\mathscr{P}(\mathrm{M}))$; in other words, the operator $\mathscr{P}$ commutes with restriction and isomorphism.

Examples of free operators are the selector, the + or − operators, the identifier, the connector, and the rank-changer (see 3.4.1 to 3.4.5); the proof is immediate.

An example of a non-free operator is the quantifier $\underset{1}{\forall}{}_1^1$, which associates with every relation $\mathrm{R}(x_1)$ the relation: $\mathrm{S}(x_1) = +$ if $\mathrm{R}(x') = +$ regardless of the value x' substituted for x_1. In fact, take the set $\{a, b\}$, with the relation

$$\mathrm{R}(a) = \mathrm{R}(b) = +$$

and the set $\{a', b'\}$, with

$$\mathrm{R}'(a') = + \quad \text{and} \quad \mathrm{R}'(b') = -,$$

and let $\mathrm{S} = \underset{1}{\forall}{}_1^1 \mathrm{R}$, and $\mathrm{S}' = \underset{1}{\forall}{}_1^1 \mathrm{R}'$. Then

$$\mathrm{S}(a) = \mathrm{S}(b) = + \quad \text{and} \quad \mathrm{S}'(a') = \mathrm{S}'(b') = -.$$

The bijective mapping of $\{a\}$ onto $\{a'\}$ is a local isomorphism of R onto R′ but not of S onto S′.

The quantifier $\underset{1}{\forall}{}_1^0$ itself is not free; indeed, define relations $\mathrm{R}(x)$ over $\{a\}$ and $\mathrm{R}'(x)$ over $\{a'\}$ by $\mathrm{R}(a) = +$ and $\mathrm{R}'(a') = -$; then $\mathrm{S} = \underset{1}{\forall}{}_1^0 \mathrm{R}$ is the 0-ary relation $(\{a\}, +)$ and $\mathrm{S}' = \underset{1}{\forall}{}_1^0 \mathrm{R}'$ is $(\{a'\}, -)$. The empty mapping is a local isomorphism of R onto R′ but not of S onto S′.

Note that every quantifier $\mathscr{U}$ commutes with every bijective mapping f defined on the entire base of M; i.e., $\mathscr{U}(f(\mathrm{M})) = f(\mathscr{U}(\mathrm{M}))$. In general, however, it does not commute with restriction: in the first of the above examples, $(\underset{1}{\forall}{}_1^1 \mathrm{R}') \mid \{a'\}$ takes the value − (for the element a'), whereas $\underset{1}{\forall}{}_1^1 (\mathrm{R}' \mid \{a'\})$ takes the value +.

4.3.1. It follows from 4.1.4 that $\mathscr{P}$ is free if and only if, for any multirelations M and M′, every local isomorphism of M onto M′ *defined over at most n* elements (n = arity of $\mathscr{P}$) is a local isomorphism of $\mathscr{P}$(M) onto $\mathscr{P}$(M′).

Another theorem, involving systems $(\mathrm{M}; x_1, \ldots, x_n)$ formed by a μ-ary multirelation M and an n-tuple $(x_1, \ldots, x_n)$ over the base of M (n = arity of $\mathscr{P}$), is:

For $n \geqslant 1$, the following condition is necessary and sufficient for $\mathscr{P}$ to be free: *given arbitrary systems* $(\mathrm{M}; x_1, \ldots, x_n)$ *and* $(\mathrm{M}'; x'_1, \ldots, x'_n)$, *if the function taking* x_i *to* x'_i *is a local isomorphism of* M *onto* M′, *then*

$$\mathscr{P}(\mathrm{M})\,(x_1, \ldots, x_n) = \mathscr{P}(\mathrm{M}')\,(x'_1, \ldots, x'_n).$$

▷ If $\mathscr{P}$ is free and the function f which takes x_i to $x'_i (i = 1, \ldots, n)$ maps $\mathrm{M} \mid \{x_1, \ldots, x_n\}$ onto $\mathrm{M}' \mid \{x'_1, \ldots, x'_n\}$, then this function will also transform $\mathscr{P}(\mathrm{M}) \mid \{x_1, \ldots, x_n\}$ into $\mathscr{P}(\mathrm{M}') \mid \{x'_1, \ldots, x'_n\}$, giving the required equality. Conversely, let f be a local isomorphism of M onto M′. If the condition of the assertion is satisfied, we have

$$\mathscr{P}(\mathrm{M})\,(x_1, \ldots, x_n) = \mathscr{P}(\mathrm{M}')\,(f(x_1), \ldots, f(x_n)),$$

and the analogous equalities obtained by taking all possible n-tuples $(x_1, \ldots, x_n)$ and their images under f. From all these equalities it follows that f is a local isomorphism of $\mathscr{P}$(M) onto $\mathscr{P}$(M′); hence $\mathscr{P}$ is free. ◁

For $n = 0$, the above two assertions coincide, as follows:

For $\mathscr{P}$ to be free, it is necessary and sufficient that, for any M and M′, *if the empty mapping is a local isomorphism of* M *onto* M′ (i.e., the 0-ary relations with the same index have the same value, + or −), *then* $\mathscr{P}(\mathrm{M}) = \mathscr{P}(\mathrm{M}')$.

4.3.2. Note that *if* $\mathscr{P}, \mathscr{Q}_1, \ldots, \mathscr{Q}_h$ *are free operators, then the composite operator* $\mathscr{P}(\mathscr{Q}_1, \ldots, \mathscr{Q}_h)$ *is free* (provided it exists).

4.3.3. *If two* (μ, n)*-ary free operators* $(n \geqslant 1)$ *coincide for every multirelation whose base is a subset of* $\{1, \ldots, n\}$, *then they are identical.*

▷ Let $\mathscr{P}$ and $\mathscr{P}'$ be two operators; let M be a μ-ary multirelation with base E, and F a nonempty subset of E, containing at most n elements. Let f be a bijective mapping of F onto a subset of $\{1, \ldots, n\}$. Finally, let N be the multirelation $f(\mathrm{M} \mid \mathrm{F})$. Now, since $\mathscr{P}$ and $\mathscr{P}'$ are free, f is a local isomorphism of $\mathscr{P}(\mathrm{M})$ onto $\mathscr{P}(\mathrm{N})$ and of $\mathscr{P}'(\mathrm{M})$ onto $\mathscr{P}'(\mathrm{N})$.

Now, by hypothesis, $\mathscr{P}(\mathrm{N})=\mathscr{P}'(\mathrm{N})$, and therefore the identity on F, which is just the mapping $f^{-1}f$, is a local isomorphism of $\mathscr{P}(\mathrm{M})$ onto $\mathscr{P}'(\mathrm{M})$. In other words, $\mathscr{P}(\mathrm{M}) \mid \mathrm{F}=\mathscr{P}'(\mathrm{M}) \mid \mathrm{F}$. Since this is true for every subset F containing at most n elements of E, it follows from 3.1.1 that $\mathscr{P}(\mathrm{M})=\mathscr{P}'(\mathrm{M})$; the assertion thus follows. ◁

4.3.4. *The set of free operators of given arities is finite.*

Let μ and n be given arities ($n \geqslant 1$). The set of relations of given arity, defined on $\{1, \ldots, n\}$ or on one of its subsets, is finite, therefore so is the set of μ-ary multirelations defined on these sets. Hence, there is a finite number of mappings which, with every μ-ary multirelation defined on all or a subset of $\{1, \ldots, n\}$, associate an n-ary relation over the same base. Hence the assertion follows by 4.3.3.

For $n=0$, the assertion becomes:

Let $\mathscr{P}$ be a $(\mu, 0)$-ary free operator. Let k be the ranks in $\mu=(m_1, \ldots, m_h)$ for which $m_k=0$. Then the *value* $\mathscr{P}(\mathrm{M})$ *depends only upon the* 0-*ary relations of ranks* k *in* M. This follows from 4.3.1 in the case $n=0$.

4.3.5. *A multirelation* N *is freely interpretable in* M *if and only if, for every relation* S *in* N, *there exists a free operator* $\mathscr{P}$ *such that* $\mathrm{S}=\mathscr{P}(\mathrm{M})$ (free interpretability was defined in 4.2).

▷ If there exists a free operator $\mathscr{P}$ such that $\mathrm{S}=\mathscr{P}(\mathrm{M})$, then, setting $\mathrm{M}'=\mathrm{M}$ in the definition in 4.3, we see that every local automorphism of M is also a local automorphism of S. Since this is true for every relation S in N, the multirelation N is freely interpretable in M.

Conversely, assume that N, and hence every relation S in N, is freely interpretable in M, and let n be the arity of S. With each multirelation M′ having base E′ and the same arity as M, we associate a relation S′ on E′ as follows. For every n-tuple $x'_1, \ldots, x'_n$ over E′, there are two possible cases. *Case* 1: There exists a local isomorphism of M′ onto M which is defined on the $x'_i (i=1, \ldots, n)$ and maps them onto an n-tuple $x_1, \ldots, x_n$ over E. Then we set $\mathrm{S}'(x'_1, \ldots, x'_n)=S(x_1, \ldots, x_n)$. This latter value is independent of the specific local isomorphism used, since, for every other sequence $y_1, \ldots, y_n$ obtained in the same way as $x_1, \ldots, x_n$, the function which takes x_i to $y_i (i=1, \ldots, n)$, is a local isomorphism of M, and hence by hypothesis, of S. *Case* 2: There is no local automorphism of M′ onto M defined on the x'_i; we then set $S'(x'_1, \ldots, x'_n) = +$. The operator $\mathscr{P}$ which takes every M′ to S′ takes M to S. Furthermore, $\mathscr{P}$ is free, by 4.3.1. Indeed, for any multirelation M″ and an arbitrary n-tuple $(x''_1, \ldots, x''_n)$ over its base, let S″ denote the relation $\mathscr{P}(M'')$. If the mapping taking x'_i to x''_i $(i = 1, \ldots, n)$ is a local isomorphism of M′ onto M″, then either there exists a local isomorphism of M′ onto M defined on the x'_i, and therefore a local isomorphism of M″ onto M defined on the x''_i, so that

$$S'(x'_1, \ldots, x'_n) = S''(x''_1, \ldots, x''_n),$$

or none of these isomorphisms exist, which leads to the same equality with the value $+$. ◁

4.3.6. Let $\mathscr{P}$ be a free operator transforming every m-ary relation R into an n-ary relation $\mathscr{P}(R)$. *If* $\mathscr{P}$ *is injective* (i.e., $R \neq R'$ implies $\mathscr{P}(R) \neq \mathscr{P}(R')$), *then* $n \geqslant m$ *and there exists a free operator* $\mathscr{Q}$ *such that* $\mathscr{Q}\mathscr{P}(R) = R$ *for every* m*-ary relation* R.

▷ The number of n-ary relations over a finite number p of elements is 2^{p^n}, which is a strictly increasing function of n. Hence, since the operator is injective, we must have $n \geqslant m$. We define $\mathscr{Q}$ as follows: for every n-ary relation S and every m-tuple $x_1, \ldots, x_m$ over the base of S, let F be the set of the x_i $(i = 1, \ldots, m)$. Then, if the restriction $S \mid F$ is the image under $\mathscr{P}$ of a relation R with base F, it follows by injectivity that R is unique, and we set $\mathscr{Q}(S)\ (x_1, \ldots, x_m) = R(x_1, \ldots, x_m)$. If $S \mid F$ is not the image of any relation, we set $\mathscr{Q}(S)\ (x_1, \ldots, x_m) = +$. To see that $\mathscr{Q}$ is free, take two m-tuples x_i and x'_i $(i = 1, \ldots, m)$ each of which is mapped onto the other by a local automorphism of S and, going back to the set F of the x_i, determine whether $S \mid F$ is or is not the image of a relation under $\mathscr{P}$. It follows that every local automorphism of S defined on at most m elements is a local automorphism of $\mathscr{Q}(S)$; now use 4.3.1. ◁

If $n = m$ and $\mathscr{P}$ is injective, we see that $\mathscr{P}$ defines a permutation of the set of m-ary relations over the base. Since the number of these relations over m elements is finite, there exists an integer h such that $\mathscr{P}^h(R) = R$ for every m-ary relation R over m elements, and hence, by 3.1.1, for every m-ary relation as well. We see that the inverse operator is then $\mathscr{P}^{h-1}$. *Special case*: the operator of permutation of ranks $\rho x_2 x_3 \ldots x_m x_1$, for which $h = m$.

The preceding results can be extended to multirelations of given arity μ, by considering finite sequences of μ-ary operators which transform multirelations into multirelations.

4.3.7. By 4.3.6, we see that, given two relations R, S, *if there exists a free injective operator* $\mathscr{P}$ *such that* $\mathscr{P}(R)=S$, *then* R *and* S *are mutually freely interpretable, and arity of* $S \geqslant$ *arity* of R. The converse is also true (M. Pouzet).

4.4. Compactness lemma

Let $x_1, x_2, \ldots, x_i, \ldots$ be an infinite sequence (with possible repetitions) indexed by the positive integers, M a μ-ary multirelation defined on the set of the x_i, and $\mathscr{P}$ a (μ, n)-ary operator. It is convenient to define the *value of* $\mathscr{P}$ *for the system* $(M; x_1, x_2, \ldots, x_i, \ldots)$ as the known value $\mathscr{P}(M)(x_1, \ldots, x_n)$, where the infinite sequence is truncated after its first n terms.

Consider a denumerable set of free operators $\mathscr{P}_k$, each of the same predicarity μ, but of arbitrary arities.

Assume that for every finite set of $\mathscr{P}_k$ *there exist a sequence and a multirelation which give them the value* $+$ *; then there exist a sequence and a finite or denumerable multirelation which give all the* $\mathscr{P}_k$ *the value* $+$.

$\triangleright$ By isomorphism, we may always assume that the x_i are the positive integers, indexed in increasing order and with return to a previous integer whenever x_i is a repetition of $x_j (j<i)$. With each integer k, associate a sequence $x_{k,1}, \ldots, x_{k,i}, \ldots$ and a multirelation M_k for which $\mathscr{P}_1, \mathscr{P}_2, \ldots, \mathscr{P}_k$ all take the value $+$. The $x_{k,1}$ are all equal to 1. There are infinitely many k's for which the restrictions $M_k | \{1\}$ are the same (this is so because there are only finitely many μ-ary multirelations over any finite base). Among these k, there are infinitely many for which the $x_{k,2}$ (which take the values 1 or 2) are all equal and for which, in addition, the restrictions $M_k | \{x_{k,1}, x_{k,2}\}$ are the same. Continuing in this way, we get a sequence of integers $y_1, \ldots, y_i, \ldots$ and, for each integer i, a multirelation M_i over $\{y_1, \ldots, y_i\}$, such that each M_i is a restriction of M_{i+1}. The M_i's define a multirelation M which is their common extension to the set of y_i; the sequence y_i and the multirelation M give each $\mathscr{P}_k$ the value $+$, and the assertion is thus proved. $\triangleleft$

4.5. Active or Inactive Predicate: The Operators $\mathcal{A}\mathcal{P}$ and $\mathcal{E}\mathcal{P}$; Interpolation Lemma [CRA, 1957; LYN, 1959]

4.5.1. A predicate ρ_k of an operator $\mathcal{P}(\rho_1, \ldots, \rho_k, \ldots, \rho_h)$ is said to be *inactive* if the relation $\mathcal{P}(R_1, \ldots, R_k, \ldots, R_h)$ is the same for any two multi-relations over the same base which differ only in their k-th relation. A predicate which is not inactive is said to be *active*.

Example. All predicates in an identifier or a + operator are inactive. In the selector $\mathcal{P}(\rho_1, \rho_2)$, defined by $R_1 = \mathcal{P}(R_1, R_2)$, the predicate ρ_1 is active and ρ_2 is inactive. In the connector $\bigwedge(\rho_1, \rho_2)$, both predicates are active.

4.5.2. Let $\mathcal{P}$ be an operator of predicarity $(m_1, \ldots, m_h)$ and k one of the ranks $1, \ldots, h$. We associate with $\mathcal{P}$ operators $\mathcal{A}_k\mathcal{P}$ and $\mathcal{E}_k\mathcal{P}$ defined as follows:

$$\mathcal{A}_k\mathcal{P}(R_1, \ldots, R_k, \ldots, R_h)(x_1, \ldots, x_n) = +$$

if and only if

$$\mathcal{P}(R_1, \ldots, S, \ldots, R_h)(x_1, \ldots, x_n) = +$$

for any relation S substituted for R_k, with the same base and arity as R_k. We define $\mathcal{E}_k\mathcal{P}$ similarly, replacing the words "for any S" by "for at least one S".

Example. Let $\mathcal{P}$ be the connector $\bigvee^1_{1,1}$, which transforms every pair of unary relations R_1 and R_2 into the unary relation $\bigvee^1_{1,1}(R_1, R_2)$, which is + if either $R_1(x) = +$ or $R_2(x) = +$. Then $\mathcal{A}_2\mathcal{P}$ is the selector which transforms the pair (R_1, R_2) into the relation R_1. Indeed,

$$\mathcal{A}_2\mathcal{P}(R_1, R_2)(x) = + \quad \text{if} \quad R_1(x) = + \quad \text{or} \quad S(x) = +,$$

for any unary relation S having the same base as R_1 and R_2. Now, there always exists a relation S such that $S(x) = -$, and therefore

$$\mathcal{A}_2\mathcal{P}(R_1, R_2)(x) = + \quad \text{if and only if} \quad R_1(x) = +.$$

The notation $\mathcal{A}_k$ and $\mathcal{E}_k$ recalls the notation $\underset{i}{\forall}$ and $\underset{i}{\exists}$ for quantifiers. However, $\mathcal{A}_k$ and $\mathcal{E}_k$ are not operators but functions which transform one operator into another. Furthermore, at our level the main interest in $\mathscr{A}_k$ and $\mathscr{E}_k$ lies in 4.5.4 and 4.5.5 below. We know of no analogous interesting result for quantifiers.

4.5.3. In the operators $\mathscr{A}_k\mathscr{P}$ and $\mathscr{E}_k\mathscr{P}$, the predicate ρ_k (i.e., the predicate of rank k) is inactive.

If a predicate ρ is inactive in $\mathscr{P}$, it is also inactive in $\mathscr{A}_k\mathscr{P}$ and in $\mathscr{E}_k\mathscr{P}$. We have the deductions $\mathscr{A}_k\mathscr{P} \vdash \mathscr{P} \vdash \mathscr{E}_k\mathscr{P}$; if ρ_k is inactive in $\mathscr{P}$, then $\mathscr{A}_k\mathscr{P} = \mathscr{E}_k\mathscr{P} = \mathscr{P}$ and converseley (deducibility is defined for relations in 3.3, and for operators in 3.5.1).

If $\mathscr{P} \vdash \mathscr{Q}$, then $\mathscr{A}_k\mathscr{Q} \vdash \mathscr{A}_k\mathscr{Q}$ and $\mathscr{E}_k\mathscr{P} \vdash \mathscr{E}_k\mathscr{Q}$.

If $\mathscr{P} \vdash \mathscr{Q}$ and ρ_k is inactive in $\mathscr{P}$, then $\mathscr{P} \vdash \mathscr{A}_k\mathscr{Q}$; if ρ_l is inactive in $\mathscr{Q}$, then $\mathscr{E}_l\mathscr{P} \vdash \mathscr{Q}$. If ρ_k is inactive in $\mathscr{P}$ and ρ_l is inactive in $\mathscr{Q}$, then

$$\mathscr{P} \vdash \mathscr{E}_l\mathscr{P} \vdash \mathscr{A}_k\mathscr{Q} \vdash \mathscr{Q}.$$

4.5.4. *If the operator $\mathscr{P}$ is free, then $\mathscr{A}_k\mathscr{P}$ and $\mathscr{E}_k\mathscr{P}$ are free.*

▷ Let $(\mathrm{M}; x_1, \ldots, x_n)$ and $(\mathrm{M}'; x'_1, \ldots, x'_n)$ be two systems such that the function f which takes x_i to $x'_i (i = 1, \ldots, n)$ is a local isomorphism of M onto M′. Let

$$\mathrm{M} = (\mathrm{R}_1, \ldots, \mathrm{R}_k, \ldots, \mathrm{R}_h) \quad \text{and} \quad \mathrm{M}' = (\mathrm{R}'_1, \ldots, \mathrm{R}'_k, \ldots, \mathrm{R}'_h).$$

By 4.3.1, it suffices to show that

$$\mathscr{A}_k\mathscr{P}(\mathrm{M})\,(x_1, \ldots, x_n) = \mathscr{A}_k\mathscr{P}(\mathrm{M}')\,(x'_1, \ldots, x'_n).$$

To this end, let us assume the first member to be equal to +. Then, for every m_k-ary relation S substituted for R_k, we have

$$\mathscr{P}(\mathrm{R}_1, \ldots, \mathrm{S}, \ldots, \mathrm{R}_h)\,(x_1, \ldots, x_n) = +.$$

Substitute for the relation R'_k of M′ an arbitrary m_k-ary relation S′ with the same arity and base. Then let S be an extension, over the base of M, of $f^{-1}(\mathrm{S}' \mid \{x'_1, \ldots, x'_n\})$; then f is a local isomorphism of $(\mathrm{R}_1, \ldots, \mathrm{S}, \ldots, \mathrm{R}_h)$ onto $(\mathrm{R}'_1, \ldots, \mathrm{S}', \ldots, \mathrm{R}'_h)$ and, consequently,

$$\mathscr{P}(\mathrm{R}'_1, \ldots, \mathrm{S}', \ldots, \mathrm{R}'_h)\,(x'_1, \ldots, x'_n)$$

is equal to

$$\mathscr{P}(\mathrm{R}_1,\ldots,\mathrm{S},\ldots,\mathrm{R}_h)\,(x_1,\ldots,x_n)=+\,.$$

It follows that $\not{\mathscr{P}}_k\mathscr{P}(\mathrm{M}')\,(x'_1,\ldots,x'_n)=+$. Interchanging M and M′, x_i and x'_i, we see that, conversely,

$$\not{\mathscr{P}}_k\mathscr{P}(\mathrm{M}')\,(x'_1,\ldots,x'_n)=+ \quad \text{implies} \quad \not{\mathscr{P}}_k\mathscr{P}(\mathrm{M})\,(x_1,\ldots,x_n)=+\,.$$

The assertion is thus proved for $\not{\mathscr{P}}_k\mathscr{P}$. The proof for $\mathscr{S}_k\mathscr{P}$ is the same, with + replaced by −. ◁

4.5.5. *Interpolation Lemma for Free Operators*

If $\mathscr{P}$ and $\mathscr{Q}$ are free and $\mathscr{P}\vdash\mathscr{Q}$, there exists a free operator $\mathscr{R}$ such that $\mathscr{P}\vdash\mathscr{R}\vdash\mathscr{Q}$, and the only active predicates in $\mathscr{R}$ are those active in both $\mathscr{P}$ and $\mathscr{Q}$.

Indeed, if $k, k',\ldots$ are the ranks of the inactive predicates in $\mathscr{P}$ and $l, l',\ldots$ the ranks of the inactive predicates in $\mathscr{Q}$, we have

$$\mathscr{P}\vdash\mathscr{S}_l\mathscr{S}_{l'}\ldots\mathscr{P}\vdash\not{\mathscr{P}}_k\not{\mathscr{P}}_{k'}\ldots\mathscr{Q}\vdash\mathscr{Q}.$$

Now let $\mathscr{R}$ be one of the two intermediate operators in this sequence of deductions.

4.6. Free formula, arity, predicate, value; thesis, antithesis, deduction; image of a free formula under a mapping

A formula in the sense of 1.2 is said to be *free* if its terms are:

(a) *free operators* of the same predicarity, with weight 0;

(b) *connections*, each of weight equal to its arity.

Let A be a free formula. The *predicarity of* A is the common predicarity $\mu=(m_1,\ldots,m_h)$ of the free operators in A; the *predicates of* A are the corresponding predicates $\rho_1,\ldots,\rho_h$. The arities of different free operators in A may be different; their maximum n is called the *arity of* A. Finally, μ and n are called the *arities* of A.

4.6.1. With any free formula A, we associate the *free operator represented by* A, which we denote by $\mathscr{P}_\mathrm{A}$. Every μ-ary multirelation M is said to be *assignable* to A, and is transformed into an n-ary relation $\mathscr{P}_\mathrm{A}(\mathrm{M})$ (or simply A (M)) defined as follows:

(a) if A is a free operator $\mathscr{P}$, then $A(M)=\mathscr{P}(M)$;

(b) if $A=\alpha A_1 \ldots A_h$, where α is an h-ary connection and $A_1, \ldots, A_h$ are the immediate subformulas of A, then if the images $A_1(M), \ldots, A_h(M)$ are of arities $m_1, \ldots, m_h$, the relation A(M) is $\alpha^m_{m_1, \ldots, m_h}(A_1(M), \ldots, A_h(M))$, where m is the maximum of $m_1, \ldots, m_h$ (see 3.4.4). We see that the arity of A(M) is the maximum of the arities of $A_1(M), \ldots, A_h(M)$, that is to say, the maximum of the arities of the free operators in A; this is indeed the arity of A as defined above. Since operator $\mathscr{P}_A$ is obtained by composition of free operators (those in A, plus connectors, each of which corresponds to an occurrence of a connection in A), it follows from 4.3.2 that $\mathscr{P}_A$ is free.

Let

$$A(M)(x_1, \ldots, x_n)=\mathscr{P}_A(M)(x_1, \ldots, x_n)$$

denote the value (+ *or* −) *taken by* A for M and $x_1, \ldots, x_n$, or the value of the relation A(M) for the sequence $x_1, \ldots, x_n$ over the base of M.

Examples. The formula $(\rho xz \wedge \rho zy) \vee (\rho yz \wedge \rho zx)$ transforms every binary relation R into a ternary relation S. If R is an ordering, S is the relation "z lies between x and y".

The formula $\rho xy \wedge \neg(x \equiv y)$ transforms every binary relation R into another binary relation S. If R is an ordering, S is the corresponding strict ordering, already considered in 4.2.

For every free operator $\mathscr{P}$ there evidently exists a free formula which represents $\mathscr{P}$: the formula formed by $\mathscr{P}$ itself. At the same time, there are infinitely many others. For example:

$$\mathscr{P} \wedge \mathscr{P}, \qquad \mathscr{P} \wedge (\mathscr{Q} \vee \neg \mathscr{Q}),$$

where $\mathscr{Q}$ is any operator of the same predicarity as $\mathscr{P}$ and arity less than or equal to that of $\mathscr{P}$.

4.6.2. *For every* (μ, n)*-ary free operator* $\mathscr{P}$ *with* $n \geqslant 1$ *and nonempty sequence* μ, *there exists a free formula representing* $\mathscr{P}$, *whose only free operators are the identifiers and rank-changers (each in composition with a selector). If* μ *is empty, the assertion still holds for* $n \geqslant 2$.

▷ Given a μ-ary multirelation M and an n-tuple $(x_1, \ldots, x_n)$, it follows from 4.3.1 that the value

$$\mathscr{P}(M)(x_1, \ldots, x_n)=+ \quad \text{or} \quad -$$

is dependent only upon the following factors:

(a) the equalities and inequalities among the $x_i (i=1,\dots, n)$; i.e., the values taken by each identifier $(x^i \equiv x^j)^n_\mu$, where $i, j=1,\dots, n$, $i \neq j$;

(b) the values taken by each relation R_k in M for each sequence of l terms over $x_1,\dots, x_n$, where l is the arity of R_k; i.e., the values taken by each rank-changer $(\rho_k x^{m_1}\dots x^{m_l})^n$ (in composition with the selector extracting the relation of rank k from M); $m_1,\dots, m_l=1,\dots, n$.

Let $\mathscr{P}_1,\dots, \mathscr{P}_u$ denote these identifiers and rank-changers. We see that there exists a μ-ary connection α such that $\mathscr{R}$ is represented by the formula $\alpha\mathscr{P}_1\dots\mathscr{P}_u$. ◁

We could have used only identifiers $(x^i \equiv x^j)^m$ for which $m=j$ and rank-changers $(\rho_k x^{m_1}\dots x^{m_l})^m$ for which $m=\max(m_1,\dots, m_l)$; the above assertion would then be modified as follows:

There exists a free formula representing $\mathscr{P}$ whose only free operators are:

(a) *identifiers* $(x^i \equiv x^j)^j$, $j>i$;

(b) *rank-changers* $(\rho_k x^{m_1}\dots x^{m_l})^{\max(m_1,\dots, m_l)}$;

(c) *operators* $+^n$.

It suffices to replace

$$(x^i \equiv x^j)^n \quad \text{by} \quad (x^i \equiv x^j)^j \wedge +^n$$

and

$$(\rho_k x^{m_1}\dots x^{m_l})^n \quad \text{by} \quad (\rho_k x^{m_1}\dots x^{m_l})^{\max(m_1,\dots, m_l)} \wedge +^n.$$

In the case of arity $n=0$, the value $\mathscr{P}(M)$ depends only upon the values of the 0-ary relations in M (if these exist), so that $\mathscr{P}$ is representable by a free formula whose only operators are selectors (extracting a 0-ary relation from M) and sometimes the operator $+^0_\mu$; the previous assertion still holds, in the second version.

In the case in which μ *is empty and* $n=1$, we have neither identifier nor rank-changer; the $+$ operator is still necessary.

4.6.3. A free formula A of arities (μ, n) is a *thesis* if $A(M)(x_1,\dots, x_n)=+$ for every μ-ary multirelation M and every n-tuple $(x_1,\dots, x_n)$ over its base. We write $\vdash A$. The formula is called an *antithesis* if $A(M)(x_1,\dots, x_n)=-$ for every M and $(x_1,\dots, x_n)$; we then write $A\vdash$. If the base is empty, the only condition on a thesis is that it not take the value $-$.

Consider a connective formula which is a tautology (2.2.2), and replace each atom, at each of its occurrences, by a free formula; we get a free formula which we again call a *tautology*. The definition of a free formula which is a *contradiction* is analogous.

Every tautology is a thesis and every contradiction is an antithesis, but the converse does not hold. For example, $\rho x^1 \Rightarrow \rho x^1$ is a tautology; $(x^1 \equiv x^2 \wedge \rho x^1) \Rightarrow \rho x^2$ is a thesis but not a tautology.

A free formula which is not an antithesis, in other words, which takes the value + for a multirelation and a sequence of elements from the base of the multirelation, is said to be *consistent*. Every thesis is consistent, but the converse does not hold; for example, ρx^1 is consistent but is not a thesis (i.e., it is neither a thesis nor an antithesis).

By 4.3.3, *a sufficient condition for a* (μ, n)*-ary free formula to be a thesis is that it take the value + for all n-tuples* $(x_1, \ldots, x_n)$ *over* $\{1, \ldots, n\}$ *and all* μ*-ary multirelations with base* $\{x_1, \ldots, x_n\}$.

We can even limit ourselves to *the n-tuples in which*

$$x_1 = 1, \; x_2 \leqslant 2, \; \ldots, \; x_n \leqslant n.$$

Since these n-tuples and the multirelations are finite in number, we have here an *algorithm which determines whether a given free formula is a thesis or not.*

Example. To show that $(x^1 \equiv x^2 \wedge \rho x^1) \Rightarrow \rho x^2$ is a thesis, first consider the sequence $x_1 = x_2 = 1$, assigning to the predicate ρ first the relation $R(1) = +$ and then the relation $R'(1) = -$. In each case, the formula takes the value +. Next, consider the sequence $x_1 = 1$, $x_2 = 2$; then the identifier $x^1 \equiv x^2$ takes the value $-$ and the formula takes the value +, for any relation R with base $\{1, 2\}$ assigned to the predicate ρ.

We leave it to the reader to formulate an algorithm which determines whether a formula A of arity 0 (all free operators of arity 0) is a thesis or not.

4.6.4. Let A and B be two free formulas of the same predicarity μ and arbitrary arities. We shall say that B *is deducible from* A, writing $A \vdash B$, if the operators $\mathscr{P}_A$ and $\mathscr{P}_B$ represented by A and B are such that $\mathscr{P}_A \vdash \mathscr{P}_B$. In other words, if n_a and n_b denote the arities of A and B, then $A \vdash B$ if, for every M of arity μ and every sequence $x_1, x_2, \ldots, x_h$, where $h = \max(n_a, n_b)$, the condition $A(M)(x_1, \ldots, x_{n_a}) = +$ implies that $B(M)(x_1, \ldots, x_{n_b}) = +$.

Example: $x^1 \equiv x^2 \wedge \rho x^1 \vdash \rho x^2$.

On the other hand, it is not true that $\rho x^2 \vdash x^1 \equiv x^2$. Indeed, consider the sequence 1, 2 and a relation R over $\{1, 2\}$ such that $R(2) = +$, with the value of $R(1)$ arbitrary; then, for the system (R; 1, 2), the operator ρx^2 takes the value $+$, while the operator $x^1 \equiv x^2$ takes the value $-$.

The deducibility relation $\vdash$ is a pre-ordering on the set of free formulas of given predicarity. A free formula is a thesis if and only if it is deducible from every free formula, and an antithesis if every formula is deducible from it. $A \vdash B$ holds if and only if the formula $A \Rightarrow B$ is a thesis.

If $A \vdash B$ and $B \vdash A$, we say that A and B are *equideducible* and write $A \dashv\vdash B$. Equideducibility is an equivalence relation; the represented operators $\mathscr{P}_A$ and $\mathscr{P}_B$ are identical if and only if $A \dashv\vdash B$ and moreover A and B have the same arity.

Let us associate with each h-ary connection α the *operation* α which transforms every sequence of h free formulas $A_1, \ldots, A_h$ with the same predicarity into the free formula $\alpha A_1 \ldots A_h$; the operation α is compatible with equideducibility. The operations of negation, conjunction and disjunction make the pre-ordering $\vdash$ into a Boolean lattice [Boolean algebra] over the free formulas. For every free formula, there exists another which contains only the connections $\neg, \bigwedge$ (or $\neg, \bigvee$), has the same free operators, and represents the same operator.

4.6.5. Let A be a free formula, n its arity and f a mapping defined on the integers $1, \ldots, n$ and having integer values. We define the *transform* $A]f$ as the free formula obtained from A by replacing each free operator $\mathscr{A}$ of arity $p \leqslant n$ by the free operator $f_p^{p'} \mathscr{A}$, where $p' = \max(f(1), \ldots, f(p))$. We see that if $\mathscr{P}$ is the operator represented by A and n the arity of A (and hence of $\mathscr{P}$), then the *operator represented by* $A]f$ *is* $f_n^{n'} \mathscr{P}$, where $n' = \max(f(1), \ldots, f(n))$.

The mapping f will often be defined by a finite sequence of ordered pairs of integers $u' = f(u)$, $v' = f(v), \ldots$, where $u, v, \ldots$ are all distinct. We then set $f(i) = i$ for $i \neq u, v, \ldots, i \leqslant n$ and f will be denoted by $\left|{}^{u'}_{u}, {}^{v'}_{v}, \ldots\right.$; thus the transform $A]f$ will be denoted by $A]^{u'v'}_{u\,v} \ldots$.

Example. Let A be the formula

$$(\rho x^1 \bigwedge \rho x^2) \Rightarrow (x^1 \equiv x^2),$$

then $A]^1_2$ is $(\rho x^1 \bigwedge \rho x^1) \Rightarrow +^1$ (the identifier $x^1 \equiv x^2$ becomes the operator $+$).

$A]^{31}_{12}$ is $(\rho x^3 \wedge (\rho x^1)^3) \Rightarrow (x^1 \equiv x^3)$; for ρx^2 denotes the operator $(\rho x^2)^2$ and the maximum of $f(1)=3$, $f(2)=1$ is 3, so that the transform must be $(\rho x^1)^3$ and not ρx^1, which would be the abbreviation of $(\rho x^1)^1$.

If A *is a thesis, then the transform* A] *f is also a thesis.* Indeed, the operator $\mathscr{P}$ represented by A always takes the value +; for all M and all $x_1, \ldots, x_{n'}$, we have

$$f^{n'}_n \mathscr{P}(\mathrm{M})(x_1, \ldots, x_{n'}) = \mathscr{P}(\mathrm{M})(x_{f(1)}, \ldots, x_{f(n)}) = + .$$

It is also clear that *if* $\mathrm{A} \vdash \mathrm{B}$ *then* $\mathrm{A}]f \vdash \mathrm{B}]f$.

4.7. Completeness Lemma for Free Theses

Let $\mathscr{A}$ denote the set of free formulas of given predicarity μ in which the only operators are identifiers, rank-changers and + operators. We saw in 4.6.2 that every free operator is represented by a formula of $\mathscr{A}$; furthermore, if f is a mapping from integers to integers, and a formula A contains only the above-mentioned operators, then the same holds for its transform $\mathrm{A}]f$. Let $\mathscr{U}$ denote the smallest subset of $\mathscr{A}$ which contains the following formulas:

(1) every tautology of predicarity μ;

(2) every formula which reduces to a + operator;

(3) the formula $(x^i \equiv x^j \wedge x^j \equiv x^k) \Rightarrow (x^i \equiv x^k)$ for all integers $i<j<k$;

(4) the formula $x^i \equiv x^j \Rightarrow (\mathrm{A} \Leftrightarrow \mathrm{A}]^j_i)$ for all integers $i<j$ and every free formula A, where the symbol $|^j_i$ denotes the mapping $f(i)=j$, $f(x)=x$ for $x \neq i$ (see 4.6.5);

(5) if $\mathrm{A} \in \mathscr{U}$ and $\mathrm{A} \Rightarrow \mathrm{B} \in \mathscr{U}$, then $\mathrm{B} \in \mathscr{U}$ (rule of detachment).

Lemma. *The set of theses in the set* $\mathscr{A}$ *is* $\mathscr{U}$.

▷ It is clear that *every formula in* $\mathscr{U}$ *is a thesis.* We must show, conversely, that every thesis having the indicated form belongs to $\mathscr{U}$, i.e., $\mathscr{U}$ is the *complete* set of these theses.

Note first that $\mathscr{U}$ is closed under conjunction, i.e., if $\mathrm{A} \in \mathscr{U}$ and $\mathrm{B} \in \mathscr{U}$, then $\mathrm{A} \wedge \mathrm{B} \in \mathscr{U}$. Indeed, $\mathrm{A} \Rightarrow (\mathrm{B} \Rightarrow (\mathrm{A} \wedge \mathrm{B})) \in \mathscr{U}$ (since this formula is a tautology); hence, by the rule of detachment (5) we get first $\mathrm{B} \Rightarrow (\mathrm{A} \wedge \mathrm{B}) \in \mathscr{U}$ and then $\mathrm{A} \wedge \mathrm{B} \in \mathscr{U}$. Similarly, one shows that if $\mathrm{A} \in \mathscr{U}$ then $\mathrm{A} \vee \mathrm{B} \in \mathscr{U}$ and $\mathrm{B} \Rightarrow \mathrm{A} \in \mathscr{U}$ for every free formula B whose only operators are identifiers, rank-changers and + operators.

Let A be a free thesis. We may always assume that A is not a tautology, for otherwise $A \in \mathscr{U}$ by hypothesis.

We may assume that A has been obtained from a connective formula B by substitution of free operators for the atoms. Let B′ be the conjunctive formula equivalent to B (see 2.3.3), and A′ the free thesis obtained from B′ by the above substitution. Then $A \Rightarrow A'$ and $A' \Rightarrow A$ are tautologies, and it will suffice to show that $A' \in \mathscr{U}$ in order to deduce that $A \in \mathscr{U}$. Now A′ is of the form $((A_1 \bigwedge A_2) \bigwedge \ldots) \bigwedge A_h$, where each A_i is a free thesis, a disjunction of identifiers, rank-changers and + operators with or without the connection $\neg$. By the preceding reasoning, it will suffice to show that $A_i \in \mathscr{U}$ for $i = 1, \ldots, h$.

In each of the A_i, which we shall henceforth write simply as A, we eliminate the + operators as follows. If at least one such operator is not preceded by $\neg$, A can be reduced (with the aid of tautologies) to the form $+ \bigvee A'$; but the operator + belongs to $\mathscr{U}$ (condition (2) of 4.7), and hence also $A = (+ \bigvee A') \in \mathscr{U}$. On the other hand, if all the + operators are preceded by $\neg$, we can remove them, since if $\neg + \bigvee A'$ is a thesis then so is A′; it will thus suffice to show that $A' \in \mathscr{U}$.

It will thus suffice to consider the case in which A is a disjunction of identifiers $x^i \equiv x^j (i<j)$ and rank-changers, possibly preceded by $\neg$. Note that, by the preceding arguments, if the disjunction of h free formulas $B_1, \ldots, B_h$ is a tautology and $B_r \Rightarrow A \in \mathscr{U}$ for $r = 1, \ldots, h$, then $A \in \mathscr{U}$. Now let n be the arity of A. Let the formulas B_r all be conjunctions of identifiers $x^i \equiv x^j$, $i<j \leqslant n$, with or without the connection $\neg$. Everything is thus reduced to the case of a free thesis of the form $B \Rightarrow A$; moreover, by using suitable tautologies we can transfer to B all the identifiers which figure in A. First assume that B contains both $x^i \equiv x^j$ and $\neg x^i \equiv x^j$, or that there are three integers $i<j<k \leqslant n$ such that B contains the identifiers $x^i \equiv x^j$, $x^j \equiv x^k$ and $\neg x^i \equiv x^k$. Then, by condition (3) we get $\neg B \in \mathscr{U}$, and hence $(\neg B \bigvee A) \in \mathscr{U}$ or $(B \Rightarrow A) \in \mathscr{U}$. The only remaining possibility is that the relation defined by setting $i \sim j$ if $x^i \equiv x^j$ figures in B is an equivalence relation. It partitions the integers $1, \ldots, n$ into equivalence classes. Let f be the mapping defined on $\{1, \ldots, n\}$ which associates with each i the largest integer in the class of i. By condition (4), we have $B \Rightarrow (A \Leftrightarrow A]f) \in \mathscr{U}$. Since this last formula and $B \Rightarrow A$ are both theses, $B \Rightarrow A]f$ must also be a thesis. Thus the disjunction $A]f$ contains two occurrences of the same rank-changer, both with and without $\neg$; otherwise, there would exist a relation (assignable to the predicate) and a sequence $x_1, \ldots, x_n$ over its base,

which give B the value + and $A]f$ the value − (take the same value for x_i and x_j when $x^i \equiv x^j$ figures in B, and distinct values when $\neg x^i \equiv x^j$ figures in B, and assign a relation R to the predicate ρ with $R(x_{m_1}, \ldots, x_{m_k})$ equal to − when $\rho x^{m_1} \ldots x^{m_k}$ figures in A and to + when $\neg \rho x^{m_1} \ldots x^{m_k}$ figures in A). It thus follows that $A]f$ is a tautology, whence $A]f \in \mathscr{U}$, and consequently $B \Rightarrow A]f \in \mathscr{U}$; since $B \Rightarrow (A \Leftrightarrow A]f) \in \mathscr{U}$, suitable tautologies now show that $B \Rightarrow A \in \mathscr{U}$. ◁

4.8. UNIVERSAL CLASS

A class $\mathscr{A}$ of multirelations of the same arity is said to be *universal* [TAR, 1953] if there exists a free operator $\mathscr{P}$ such that R is in $\mathscr{A}$ if and only if its image $\mathscr{P}(R)$ never takes the value − (if the base is nonempty, an equivalent condition is that $\mathscr{P}(R)$ always takes the value +). The operator $\mathscr{P}$, or any free formula representing $\mathscr{P}$, is said to *represent* the class $\mathscr{A}$.

Examples of universal classes. The empty class, represented by the operator $-^0$; the class of all multirelations of given arity, represented by a thesis. The class of reflexive binary relations, represented by the free formula ρxx; indeed, this formula transforms every binary relation R into a unary relation S over the same base, which always takes the value + or takes − for an element of the base according to whether R is reflexive or not. The class of transitive binary relations, represented by $(\rho xy \wedge \rho yz) \Rightarrow \rho xz$. The class of pre-orderings, represented by the conjunction of the above two free formulas. The classes of equivalence relations, orderings, chains (= total orderings). For a given integer p, the class of multirelations of fixed arity defined over at most p elements: the representing formula is the disjunction of the formulas $x^i \equiv x^j$ for $i<j$, $1 \leqslant i, j \leqslant p+1$. For $p=0$, we get the class which consists of the single multirelation with empty base: take the free operator $-^1$, which maps every multirelation with base E onto the unary relation which is identically equal to − on E; when E is empty, this gives the unary relation with empty base, whose value is neither + nor −.

The family of universal classes is denumerable (for the set of formulas is denumerable, and the above examples alone yield denumerably many classes).

4.8.1. *The intersection of two universal classes is a universal class.* To prove this, take the conjunction of the formulas that represent each class. *The union is also universal*, but here it is not sufficient to consider the disjunction of the two representing formulas: one must first (if necessary) carry out a bijective change of indices (this does not alter the universal class represented by the formula), in such a way that the free operators of the first formula and those of the second have no active index in common. For example, $x^1 \equiv x^2$ and $x^1 \equiv x^3 \vee x^2 \equiv x^3$ each represent the class of multirelations whose base contains a single element, but their disjunction represents the class of multirelations whose base contains one or two elements. To get back the original classes, it suffices to consider the disjunction of $x^1 \equiv x^2$ and $x^3 \equiv x^5 \vee x^4 \equiv x^5$.

If $\mathscr{A}$ is a universal class and R *belongs to $\mathscr{A}$, then any multirelation which is embeddable in* R *belongs to $\mathscr{A}$.* In other words, every universal class is closed under embeddability (see 3.2.3).

▷ Let A be a free formula representing $\mathscr{A}$. If A(R) is a relation which is always equal to $+$; then, for any subset F of the base of R, the free formula A transforms R | F into A(R) | F, which is always $+$. ◁

Given an integer p, the class of multirelations of fixed arity defined on a base containing more than p elements is not universal, since it is not closed under embeddability. Thus, *the complement of the universal class of multirelations defined on $\leqslant p$ elements* (see the above examples) *is not a universal class.*

4.8.2. Let $\mathscr{A}$ be a universal class of μ-ary multirelations, represented by a free formula A of arities (μ, n).

A multirelation R *belongs to $\mathscr{A}$ if and only if every restriction of* R *to* $0, 1, \ldots, n$ *elements belongs to $\mathscr{A}$* [VAU, 1953].

▷ If R is in $\mathscr{A}$, the same holds for its restrictions, as proved above. Conversely, if all restrictions of R to at most n elements belong to $\mathscr{A}$, then every restriction of A(R) to at most n elements always has the value $+$; thus A(R) always has the value $+$. ◁

It follows that the class of ordinals and their isomorphic images (chains for which every restriction has a minimal element) is not a universal class, since it defines the same finite restrictions as the class of all chains. Thus a class of relations may be closed under embeddability (isomorphism and restriction) and yet not be universal.

4.8.3. *Let* U *be a multirelation with finite base. Then the class of multirelations with the same arity in which* U *is not embeddable is universal.*

▷ We prove the statement for an m-ary relation U defined on the integers $1, \ldots, p$.

U is not embeddable in an m-ary relation R if and only if, for any sequence of p elements $a_1, \ldots, a_p$ of the base of R, either two of them are identical, or they are all distinct and the bijective mapping taking i to $a_i (i=1, 2, \ldots, p)$ does not transform U into a restriction of R. With every sequence of m integers $i_1, \ldots, i_m$ from 1 to p, we associate the free formula

$$(\rho x^{i_1} \ldots x^{i_m})' = \neg \rho x^{i_1} \ldots x^{i_m} \quad \text{or} \quad \rho x^{i_1} \ldots x^{i_m}$$

according to whether $U(i_1, \ldots, i_m) = +$ or $-$. The required condition is equivalent to the statement that the disjunction of the formulas $x^i \equiv x^j$ $(i<j, 1 \leqslant i, j \leqslant p)$ and $(\rho x^{i_1} \ldots x^{i_m})'$ $(i_1, \ldots, i_m$ integers between 1 and $p)$ takes the value $+$ for every sequence of p elements from the base of R. ◁

4.8.4. *A class $\mathscr{A}$ of multirelations of the same arity is universal if and only if there exist finitely many multirelations* $U_1, \ldots, U_h$ *over finite bases such that* R *is in $\mathscr{A}$ if and only if none of* $U_1, \ldots, U_h$ *is embeddable in* R.

▷ If such U's exist, then $\mathscr{A}$ is universal by virtue of 4.8.3 and the fact that the intersection of a finite number of universal classes is universal.

Conversely, if $\mathscr{A}$ is universal, let U denote any multirelation not in $\mathscr{A}$ but such that all its proper restrictions are in $\mathscr{A}$. Up to isomorphism, there are only finitely many such multirelations; in fact, if n is the arity of a free formula representing $\mathscr{A}$, then, by 4.8.2, the base of each U contains n elements or less. If R is in $\mathscr{A}$, then no U is embeddable in R since $\mathscr{A}$ is closed under embeddability. On the other hand, if R is not in $\mathscr{A}$, there exists at least one restriction of R to $\leqslant n$ elements which does not belong to $\mathscr{A}$. Consider some such restriction for which the number of elements is minimal: all its proper restrictions are in $\mathscr{A}$, and so it must be one of the U's (up to isomorphism), so that one of the U's is embeddable in R. ◁

A necessary and sufficient condition for $\mathscr{A}$ to be universal is that there exist an integer n with the following property: R *belongs to $\mathscr{A}$ if and only if every restriction of* R *to* $0, 1, \ldots, n$ *elements belongs to $\mathscr{A}$* [VAU, 1953]. This follows from 4.8.2 and 4.8.4, as can be seen by letting the U's be the multirelations not in $\mathscr{A}$ whose proper restrictions are in $\mathscr{A}$.

As examples of this theorem, consider the universal classes given above. The multirelations U for the empty class are the multirelations with empty base. For the class of all multirelations of given arity, we take the empty set of U's. For reflexive binary relations, take a single multirelation U over the base consisting of the single integer 1, defined by $U(1, 1) = -$. For symmetric binary relations, take all multirelations U over the integers 1, 2 such that $U(1, 2) = +$ and $U(2, 1) = -$. For the class of chains, take the relation U defined for 1 alone such that $U(1, 1) = -$, and then the two reflexive relations defined over 1, 2 such that $U(1, 2) = U(2, 1) = +$ or $-$, and finally the cyclic ordering over 1, 2, 3, i.e., the reflexive and antisymmetric relation such that

$$U(1, 2) = U(2, 3) = U(3, 1) = + .$$

For the class containing the multirelation with empty base, we take all multirelations over the set consisting of 1 alone.

4.8.5. We now present some consequences of 4.8.4.

(1) Two universal classes which contain the same multirelations over finite bases are identical. In fact, the sequences of multirelations U corresponding to them in the sense of 4.8.4 are the same.

(2) If a class $\mathscr{A}$ reduces to a finite number of finite multirelations of the same arity (with their isomorphic images), then $\mathscr{A}$ is universal if and only if it is closed under embeddability. To verify this, apply, e.g., the second assertion of 4.8.4. with $n = (\text{maximum cardinality of the bases}) + 1$.

(3) Given a class $\mathscr{A}$ which is closed under $\leqslant$ (in other words, closed under isomorphism and passage to restrictions) and an integer a, consider the class $\mathscr{A}'$ obtained by removing from $\mathscr{A}$ all relations with cardinalities $< a$; let $\mathscr{A}_a$ be the closure of $\mathscr{A}'$ (i.e., $\mathscr{A}'$ plus all restrictions of cardinalities $< a$). If $\mathscr{A}$ is universal, then so is $\mathscr{A}_a$. Hence we deduce that if $\mathscr{A}$ is universal, then any class closed under $\leqslant$, which coincides with $\mathscr{A}$ as regards relations of cardinality greater than a given integer, is universal.

4.8.6. Let $\mathscr{A}$ be a class of finite multirelations of the same arity, closed under embeddability. A finite multirelation U is said to be a *bound* of $\mathscr{A}$ if U is not in $\mathscr{A}$ but every proper restriction of U is in $\mathscr{A}$. By 4.8.4, such a class is a universal class restricted to finite bases if and only if it has only finitely many bounds.

A finite multirelation V is said to be an *antibound* of $\mathscr{A}$ if V is in $\mathscr{A}$ but no proper extension of V is in $\mathscr{A}$. A universal class may contain infinitely many nonisomorphic antibounds; in other words, it may contain antibounds over arbitrarily large finite bases. The following example is due to [MALI, 1967]. For every positive integer p, take the base of integers $1, \dots, p$ and the quadrirelation $M_p = (I_p, C_p, A_p, B_p)$, where I_p is the chain of integers from 1 to p, C_p is the successor relation (equal to $+$ for pairs x, y such that $y = x+1$), A_p is the singleton of 1 and B_p the singleton of p. Starting from $p=7$, it can be seen that the multirelations M_p have the same restrictions to 0, 1, 2, 3 elements (up to isomorphism). Consider the quadrirelations U over 1, 2, or 3 elements which cannot be embedded in M_7; the universal class of quadrirelations in which no U is embeddable contains M_p for every $p \geqslant 7$. We see that every proper extension of M_p ($p \geqslant 7$) has restrictions to 1, 2, or 3 elements which are not embeddable in M_p. Thus each M_p is an antibound for the class in question.

Problem. Does there exist a universal class of binary relations with infinitely many antibounds? Positive answer by G. Lopez.

4.8.7. Any intersection $\mathscr{B}$ of universal classes in some finite or infinite (hence, of course, denumerable) set is called a *δ-universal class.* Clearly, R is in $\mathscr{B}$ if and only if every finite restriction of R (up to isomorphism) is in $\mathscr{B}$. Any class which is closed under isomorphism and satisfies the above condition is a δ-universal class. If we reduce it to its finite relations, it is simply a class closed under embeddability.

Given a multirelation R, the multirelations whose finite restrictions are all embeddable in R constitute a δ-universal class $\mathscr{R}$ with the filtration property that, for any A, B $\in \mathscr{R}$, there exists C $\in \mathscr{R}$ with C $\geqslant$ A, B (use 3.2.6). Conversely, any δ-universal class with the filtration property is obtained in this way from a multirelation (see 3.2.4).

4.8.8. *If $\mathscr{A}$ is a universal class and $\mathscr{P}$ a free operator, then the inverse image $\mathscr{P}^{-1}(\mathscr{A})$ (see 3.5.6) is universal. The direct image $\mathscr{P}(\mathscr{A})$ is an intersection of universal classes.*

▷ Let S be a multirelation with base E; suppose that, for every finite subset F of E, the restriction S | F belongs to $\mathscr{P}(\mathscr{A})$. By 4.8.7, it will suffice to show that S belongs to $\mathscr{P}(\mathscr{A})$. By hypothesis, for every finite subset F there exists a nonempty set U_F of multirelations X over the base F, belonging to $\mathscr{A}$, and such that $\mathscr{P}(X) = S \mid F$. The multirelations U_F satisfy the assumptions of the coherence lemma 3.1.3; assuming the ultrafilter axiom, we see that there exists a multirelation R with base E such that $R \mid F$ is in U_F for every finite F. Thus $\mathscr{P}(R \mid F) = S \mid F$ and, since $\mathscr{P}$ is free, $(\mathscr{P}(R)) \mid F = S \mid F$ for every F, and so $\mathscr{P}(R) = S$. Moreover, every restriction $R \mid F$ belongs to $\mathscr{A}$, and therefore R is also in $\mathscr{A}$, by 4.8.2. $\lhd$

The direct image of a universal class under a free operator is not necessarily universal (unpublished result of M. Pouzet, 1971).

$\rhd$ Let $\mathscr{A}$ be the class of all (partial) orderings, which we already know to be universal, and $\mathscr{P}$ the operator represented by $\rho xy \bigvee \rho yx$. The corresponding binary relations might be termed *comparabilities* (x and y are comparable in a certain ordering). With each positive integer p, we associate the binary cycle C_p defined on the integers $1, \ldots, p$, i.e., the reflexive relation assuming the value $+$ for the pairs $(1, 2)$, $(2, 3), \ldots$, $(p-1, p)$, $(p, 1)$. It suffices to prove that, for every odd p, the cycle C_p is a bound for the class of comparabilities. Remove one element from the base, say p. The resulting restriction is the comparability relation corresponding to the ordering: 1 and 3 precede 2, then 3 and 5 precede 4, etc. On the other hand, if the cycle C_p is a comparability relation, we may assume that 1 precedes p and 2, and then of necessity 3 precedes 2 and 4, etc., and since p is odd we must put p "before" 1, so that $p=1$; contradiction. $\lhd$

If $\mathscr{P}$ is a free operator, and $\mathscr{A}$ the class of all multirelations assignable to $\mathscr{P}$, then $\mathscr{P}(\mathscr{A})$ is not always a universal class.

Take $\mathscr{P}(R)(a, b) = +$ iff $a \neq b$ and $R(a, b) = R(b, a) = +$ and $R(a, a) \neq$ $\neq R(b, b)$. Then every odd cycle is a bound (share the base into the a's such that $R(a, a) = +$ and the b's such that $R(b, b) = -$, to see that only even cycles can be obtained); result of M. Jean and M. Pouzet.

EXERCISES

1

Call a relation R with base E a *constant* if, for all elements a, b of E, the permutation (a, b) exchanging a and b and preserving all other elements of E also preserves R.

(1) Show that an n-ary relation R is constant if and only if

$$R(a_1, \ldots a_n) = R(a'_1, \ldots, a'_n)$$

whenever the sequence $a'_1, \ldots, a'_n$ is a bijective image of $a_1, \ldots, a_n$. In other words, the value $R(a_1, \ldots, a_n)$ depends only on the equalities and inequalities subsisting among the values $a_1, \ldots, a_n$. Hence deduce that R is constant if and only if: (i) all its restrictions to 1, 2, ..., n elements are isomorphic; (ii) each of these restrictions is preserved under any permutation of its base. Finally, deduce from this that R is constant if and only if every permutation of E preserves R.

(2) Assume that the base E comprises at least n elements. Show that the number of constant n-ary relations with base E depends only on n and not on E; the number in question is $2^{h(n)}$, where $h(0)=1$ and $h(n+1)=C_0^n h(0)+C_1^n h(1)+\cdots+C_n^n h(n)$.

(3) Note that every restriction of a constant relation is constant. Given the constant n-ary relation R with base E, show that for every superset E′ of E there exists a constant extension of R to the base E′. If E contains at least n elements, this extension is unique; give a counterexample for $n=2$ and E containing a single element.

Let R be constant, and S another n-ary relation. If every restriction of S to 1, 2, ..., n elements is isomorphic to a restriction of R, then S is constant; moreover, if R and S have the same base then R = S.

(4) Let R be any n-ary relation with a base E and f a permutation of E which does not preserve R. For example, let $a_1, \ldots, a_n$ be such that

$$R(a_1, \ldots, a_n) = + \quad \text{and} \quad R(f(a_1), \ldots, f(a_n)) = - .$$

By 3.2.2, we know that there exist two elements a, b of E such that $b=f(a)$ and the permutation (a, b) does not preserve R. Use this to give another proof of the statement that R is constant if and only if every permutation of E preserves R.

(5) Show that the following condition is necessary and sufficient for a relation R with base E to be constant: There exists a free formula A containing the indices 1, ..., n and built up from connections and identifiers, such that $R(x_1, \ldots, x_n) = +$ for all sequences $x_1, \ldots, x_n$ that satisfy the formula A. Construct free formulas representing the constant unary ($n=1$) and binary ($n=2$) relations.

(6) Define a universal formula to be a free formula preceded by the sequence of universal quantifiers corresponding to all its indices.

If R is constant, deduce from the foregoing that there exist a universal formula with n indices and a predicate replaceable by R which defines R on E up to isomorphism. In other words, this formula is true for R and its isomorphic images and false for all other relations over E. If E is finite, give an example of a binary relation R with base E, which is not constant but is nevertheless defined on E (up to isomorphism) by a universal formula (the number of indices may then exceed $n=2$).

(7) Let R be an n-ary relation with a denumerable base. Suppose that every relation S with denumerable base whose restriction to 1, 2, ..., n elements is isomorphic to a restriction of R must be isomorphic to R. One can then show that R is constant. This is easily done for $n=1$. Prove the assertion for $n=2$; first establish (or assume) that every binary relation

with denumerable base has a restriction with denumerable base which is either constant or a chain (total ordering) in the broad ($\leqslant$) or narrow ($<$) sense.

(8) Hence deduce (again for $n=1$ and $n=2$) the following converse of the statement in (6). If R is an n-ary relation with denumerable base E and there exists a universal formula with n indices which defines R on E, up to isomorphism, then R is constant.

2

Given an infinite set E and a finite subset F of E, a multirelation M with base E is said to be F-*finitistic* if, for any pair of elements a, b in $E-F$, the transposition exchanging a and b (and preserving every other element) is an automorphism of M. The multirelation M is said to be *finitistic* if there exists a finite subset F of its base such that M is F-finitistic.

(1) Show that M is F-finitistic if and only if every permutation of E which preserves each element of F is an automorphism M. Show that M is F-finitistic if and only if M is interpretable by the sequence of unary relations each of which is true for a unique element of F. In the unary case, a relation is finitistic if it takes the value + over a finite set or the complement of a finite set. Finally, M is finitistic if and only if each of its relations is finitistic.

(2) Show that if M is both F- and G-finitistic, it is also ($F\cap G$)-finitistic; thus there exists a minimal F, called the *kernel* of M; note that the kernel is empty if and only if M is a constant.

(3) M is finitistic, with kernel containing $\leqslant p$ elements, if and only if there exist at most p elements a of the base, with each of which one can associate at least $p+1$ elements b such that the transpositions (a, b) change M.

(4) A multirelation M with denumerable base E is finitistic if and only if the set of isomorphic images of M over the base E either contains M alone (this is the case in which the kernel is empty) or is denumerable.

(5) Two multirelations are said to be $(1, p)$-equivalent if they have the same restrictions (up to isomorphism) to $1, \ldots, p$ elements, or, alternatively, if they assign the same value to any prenex formula in which the quantifiers are either all $\forall$ or all $\exists$, with at most p bound indices. Show that M is finitistic if and only if there exists an integer p such that every multirelation over the same base which is $(1, p)$-equivalent to M is isomorphic to M.

3

A multirelation M is said to be *enchainable* if there exists a chain T over the same base such that M is freely interpretable by T (i.e., every local automorphism of T is a local automorphism of M). *Examples:* the relation identically equal to +, the identity, the cyclic chain: $A(x, y, z)=+$ if x, y, z are ordered by T in one of the orders x, y, z or y, z, x, or z, x, y; the relation: $E(x, y, z)=+$ if z lies between x and y in the ordering T. Every restriction of an enchainable multirelation is enchainable.

(1) Using the ultrafilter axiom, show that if every finite restriction of M is enchainable, then M is enchainable.

(2) Show that every multirelation with denumerable base has a denumerable enchainable restriction. *Hint:* Consider the concatenation MT, where T is a chain with the same base as M, and then use Ramsey's theorem (Chapter 3, Exercise 4), considering sets of m elements (where m is the maximum arity of M).

(3) If a multirelation M with base E is enchainable, then for every superset E′ of E there exists an enchainable extension of M to E′. If E contains at least m elements ($m=$ maximum arity) and E′ is finite, the enchainable extension is unique up to isomorphism.

(4) Assume that, given a finite sequence μ of integers, there exists an integer a such that

every μ-ary multirelation whose restrictions of cardinality a are enchainable is itself enchainable [FRAS, 1965]. Deduce that an enchainable multirelation can have only finitely many bounds (see Chapter 3, Exercise 8: A is a bound of M if A is finite, not $\leqslant$M, but every proper restriction of A is $\leqslant$M).

4

A multirelation M is said to be *p-monomorphic* (p a natural number) if the restrictions of M to p elements of its base are all isomorphic, or if the base contains less than p elements. A multirelation is *monomorphic* if it is p-monomorphic for every natural number p. For example, any chain, or, in general, enchainable multirelation (freely interpretable by a chain; see Exercise 3) is monomorphic. There are nevertheless monomorphic relations which are not enchainable: the cycle over three elements a, b, c, defined by $R(a, b) = R(b, c) = R(c, a) = +$, all other values being $-$. Note that, by Exercise 3, every multirelation with denumerable base has an enchainable, and hence monomorphic, denumerable restriction.

(1) Using the ultrafilter axiom, show that every monomorphic multirelation with infinite base is also enchainable. *Hint:* First show that every finite restriction is enchainable (use Ramsey's theorem; see Chapter 3, Exercise 4, (2)). One can also say: If M is infinite and monomorphic, there exists a chain T such that MT is monomorphic.

(2) Note that a sufficient condition for M to be p-monomorphic is that every restriction of M to $p+1$ elements be p-monomorphic. Hence deduce that the class of p-monomorphic multirelations of a given arity is universal, i.e., there exists a free formula A such that A(M) is a relation which is always + if and only if M is p-monomorphic (see 4.8).

(3) Assume the following theorem valid [FRAS, 1965]. Given a finite sequence μ of integers, there exist two integers a and p such that every p-monomorphic μ-ary multirelation over a base of cardinality $\geqslant a$ is enchainable (therefore monomorphic). Show that the class of enchainable multirelations and the class of monomorphic multirelations, of given arity, are universal (this result is due to [JEA, 1967]). *Hint:* Use 4.8.5, (3).

5

A relation R is said to be *homogeneous* if every local automorphism of R defined on a finite subset of its base can be extended to an automorphism of R. *Examples:* the relation identically equal to +, the chain (total ordering) of rationals; recall that every dense denumerable chain with no minimal or maximal element is isomorphic to the latter chain.

(1) Let E be a denumerable base; show that the relation R is homogeneous if and only if, for any finite subset F of E, any local automorphism f of R defined on F, and any element u of $E-F$, there exists a local automorphism of R which extends f and is defined on $F\cup\{u\}$.

(2) Show that for every positive integer n there exists a denumerable and homogeneous n-ary relation in which every denumerable n-ary relation is embeddable (recall that $R\leqslant S$ means that S has a restriction isomorphic to R).

(3) Show that if two n-ary relations R and R′ are denumerable, homogeneous and have the same finite restrictions (up to isomorphism), then R and R′ are isomorphic. In particular, for every n, the n-ary relation of (2) is unique up to isomorphism; the chain of rationals is the only homogeneous denumerable chain.

(4) Let R be a denumerable and homogeneous relation, S a denumerable relation whose finite restrictions are all $\leqslant$R; show that then $S\leqslant R$.

(5) Recall that the set $\mathscr{R}$ of finite restrictions of a relation R satisfies the conditions of 3.2.4:

(i) if $A\in\mathscr{R}$ and $B\leqslant A$, then $B\in\mathscr{R}$;

(ii) if A, B $\in \mathscr{R}$, there exists C $\in \mathscr{R}$ such that C $\geqslant$ A and C $\geqslant$ B, and conversely, a set $\mathscr{R}$ satisfying these conditions is the set of finite restrictions of a certain relation. Show that there exists a homogeneous and finite or denumerable relation whose finite restrictions constitute the set $\mathscr{R}$, if and only if $\mathscr{R}$ satisfies (i) and the following condition (iii) of *amalgamability*, which is stronger than (ii):

(iii) if A, B, C $\in \mathscr{R}$ and f is an isomorphism of C onto a restriction of A, g an isomorphism of C onto a restriction of B, then there exist a relation D $\in \mathscr{R}$, an isomorphism f' of A onto a restriction of D and an isomorphism g' of B onto a restriction of D, such that $f'f(x) = g'g(x)$ for every element x of the base of C.

Hence deduce that there exists a denumerable and homogeneous (partial) ordering whose restrictions are all the finite orderings. By contrast, a tree, or ordering in which all elements preceding a given element are comparable, cannot be denumerable and homogeneous and at the same time admit all finite trees as restrictions.

(6) Let us say that R is *quasi-homogeneous* if, for every finite subset F of its base, there exists a finite superset G of F such that every local automorphism of R defined on G can be extended to an automorphism of R. Define an *almost-homogeneous* relation in similar fashion: every local automorphism defined on F that can be extended to G can also be extended to an automorphism of R (this definition is due to J. F. Pabion). Show that part (3), but not part (4), of this exercise is valid for these new concepts. Show that the three concepts do not coincide.

For more information on this subject, the reader is referred to [JON, 1965] and [CAL, 1967]. There exist a continuum-power set of non isomorphic, denumerable, homogeneous relations [WAR, 1972].

6

Let R be an m-ary relation with base E; a subset D of E is called an R-*interval* if every local automorphism of the restriction R | D, when extended by the identity on E − D, becomes a local automorphism of R. Show that if R is a chain, then D is an R-interval if and only if D is closed under the "betweenness" relation: for every x and y in D, every z lying between x and y is also in D. If R is a partial ordering, D is an R-interval if and only if every element of E − D either precedes all elements of D, or follows them, or is incomparable with all of them.

(1) Derive the following logical version of the concept of an R-interval: If two m-tuples of elements of D give the same truth value to every free formula with indices $1, 2, \ldots, m$ in which R is substituted for the predicate, then they still give the same value to every free formula with indices $1, 2, \ldots, m, m+1, \ldots, m+p$ in which R is substituted for the predicate and elements of E − D for the indices $m+1, \ldots, m+p$.

(2) Show that every intersection of R-intervals is an R-interval; the union of a set of R-intervals, totally ordered by inclusion, is an R-interval. The empty set, the base, each singleton (set containing a single element of the base), are all R-intervals. Give an example of a relation for which there are no intervals other than these three types.

(3) Let D be an R-interval; for every subset E′ of the base E, consider the restriction R | E′, and show that the intersection D∩E′ is an (R | E′)-interval. In particular, if E′ contains D, then D is an (R | E′)-interval. If D is an R-interval, then any (R | D)-interval is an R-interval contained in D and also the intersection of an R-interval with D.

(4) Consider two R-intervals D and D′, and assume that for every restriction of R to at most $m-1$ elements (m = arity) there is an isomorphic restriction in R | (D∩D′). Show that the union D∪D′ is an R-interval (use 4.1.4). In the case of a chain, $m = 2$ and all restrictions to one element are isomorphic; this gives another proof of the fact that the union of two intervals with a common element is an interval.

(5) Generalize the preceding investigation: let F be an arbitrary subset of the base E, and call a set D disjoint from F an (R, F)-interval if every local automorphism of R/D, extended by the identity on F, defines a local automorphism of R. In this framework, define the classical concept of an interval in a chain, or of an ordering defined by the set F of its two extreme values.

7

Let R be an m-ary relation with base E and E* a superset of E. An extension R* of R to E* is said to be *canonic* if every local automorphism of R, extended by any bijective mapping of a subset of E* − E onto another subset, gives a local automorphism of R*. There always exists a canonic extension; for example, define an extension of R to E* which takes the value + for any m-tuple of which at least one term belongs to E* − E.

(1) Let D be a subset of E such that every restriction of R to at most $m-1$ elements has an isomorphic restriction in R | D; note that there exists a finite set D of this description. Show that if two canonic extensions of R to E* have the same restriction to the set obtained by adjoining m elements of E* − E to D, they are identical. Hence conclude that there are only finitely many canonic extension of R to E*.

(2) Let R* be a canonic extension of R to E* and let E** be a superset of E*. Show that there exists a canonic extension of R to E** which is an extension of R*.

(3) Let A be the unary relation defined as + for every element of E and − on E* − E. Show that an extension of R to E* is canonic if and only if it is freely interpretable in SA for any extension S of R to E*.

(4) Let R and R′ be two relations, with disjoint bases E and E′, respectively. A relation S over E∪E′ is called a *sum* of R and R′ if every union of a local automorphism of R and a local automorphism of R′ is a local automorphism of S. Show that if R* and R′* are canonic extensions of R and R′ to the union E∪E′, and A is the unary relation equal to + on E and − on E′, then S is a sum of R and R′ if and only if S is freely interpretable in R*R′*A.

(5) Prove that the above "addition" of relations is associative: if R, R′, R″ are relations with disjoint bases, S is a sum of R and R′, T a sum of S and R″, then there exists a sum U of R′ and R″ such that T is a sum of R and U.

8

Let R be a multirelation with base E and F a finite subset of E. R is said to be F-*enchainable* if there exists a chain T of base E − F such that every local automorphism of T, extended by the identity on F, is a local automorphism of R. R is said to be *almost-enchainable* if there exists such a finite subset F. Particular cases: finitistic multirelations (Exercise 2), enchainable multirelations (Exercise 3).

(1) Show that for every R with denumerable base E and every finite subset F of E, there exists a denumerable restriction of R which is F-enchainable. For every integer p, there exists a restriction of R which is almost-enchainable and has the same restrictions to at most p elements.

(2) Given R, associate with every integer p an integer $U_R(p)$, defined as the number of isomorphism classes of restrictions of R to p elements. Show that if R is F-enchainable the function U_R is bounded and increasing (in each isomorphism class, take a representative for which the cardinality of the intersection of the base with F is minimum). Hence deduce that U_R is increasing for every multirelation R with infinite base (this result is due to M. Pouzet).

(3) Show that R is almost-enchainable if and only if U_R is bounded. R is almost-enchain-

able if and only if the finite-age of R has only finitely many infinite sub-finite-ages. R is almost-enchainable if and only if there exists an integer p such that every multirelation which has the same restrictions (up to isomorphism) to at most p elements has the same finite restrictions.

(4) Show that an almost-enchainable multirelation has only finitely many bounds (see Exercise 3).

CHAPTER 5

FORMULA, OPERATOR, LOGICAL CLASS AND LOGICAL EQUIVALENCE; DENUMERABLE-MODEL THEOREM

5.1. Logical formula; arity, predicate; bound, prenex and canonical formulas, free and bound indices

Given a finite sequence of positive integers $i_1, \ldots, i_r$, we define the *indefinite quantifier*, denoted by $\underset{i_1,\ldots,i_r}{\forall}$, to be the set of quantifiers $\underset{i_1,\ldots,i_r}{\forall^n_m}$ in which the indices $i_1, \ldots, i_r$ are fixed, while the predicarity m and arity n range over all values compatible with these indices. That is, $n \leqslant m$ and each of the integers $n+1, \ldots, m$ is one of the indices $i_1, \ldots, i_r$ (recall that m need not be greater than the indices). The indefinite quantifier $\underset{i_1,\ldots,i_r}{\exists}$ is similarly defined.

A formula in the sense of 1.2 will be called a *logical formula* if its terms are:

(a) *free operators*, all of the same predicarity, assigned weight 0;

(b) *connections*, each assigned weight equal to its arity;

(c) *indefinite quantifiers*, assigned weight 1.

The predicarity $\mu = (m_1, \ldots, m_h)$ of the free operators appearing in a formula P will be called the *predicarity* of P, denoted by $\mu(\mathrm{P})$; the corresponding predicates will be called the *predicates of* P, denoted by $\rho_1, \ldots, \rho_h$.

The arities of the different free operators may be distinct; starting from them, we inductively define the arity of each subformula of P, and finally the *arity of* P, denoted by $n(\mathrm{P})$:

When P reduces to a single free operator, $n(\mathrm{P})$ is the arity of this free operator.

When $\mathrm{P} = \alpha \mathrm{P}_1 \ldots \mathrm{P}_h$, where α is an h-ary connection and $\mathrm{P}_1, \ldots, \mathrm{P}_h$ the immediate subformulas of P, then $n(\mathrm{P})$ is the maximum of the arities $n(\mathrm{P}_1), \ldots, n(\mathrm{P}_h)$.

When $\mathrm{P} = \underset{i_1,\ldots,i_r}{\forall} \mathrm{Q}$ or $\underset{i_1,\ldots,i_r}{\exists} \mathrm{Q}$, where Q is the immediate subformula of P, then $n(\mathrm{P})$ is the largest integer $\leqslant n(\mathrm{Q})$ distinct from all the indices

$i_1, \ldots, i_r$; in particular, we set $n(\mathrm{P})=0$ if all the integers $1, \ldots, n(\mathrm{Q})$ are indices, and $n(\mathrm{P})=n(\mathrm{Q})$ if $n(\mathrm{Q})$ is not an index.

The predicarity μ and arity n are called the *arities* of P, denoted by (μ, n) or $(m_1, \ldots, m_h; n)$.

Example. The formula $(\underset{2}{\forall} \rho x^2 x^1) \wedge \underset{1,3}{\exists} (x^2 \equiv x^3 \vee \sigma x^2)$ includes a binary predicate $\rho_1 = \rho$ and a unary predicate $\rho_2 = \sigma$; its predicarity is thus $(2, 1)$. The rank-changers $\rho x^2 x^1$, σx^2 and the identifier $x^2 \equiv x^3$ have arities 2, 2, 3, respectively; the arity of the subformula $\underset{2}{\forall} \rho x^2 x^1$ is 1, and that of $\underset{1,3}{\exists} (x^2 \equiv x^3 \vee \sigma x^2)$ is 2; thus the arity of the formula is 2.

When P contains no quantifiers, the definition reduces to that of a free formula (see 4.6).

When the arity is 0, the formula is said to be *bound*; an example is the above formula, preceded by the quantifier $\underset{1,2}{\forall}$.

5.1.1. A formula P is said to be *prenex* if each quantifier dominates each connection, in the sense of 1.2.6. In other words, P is formed from a sequence U of quantifiers followed by a free formula A, so that P = UA. The sequence U will be called the *prefix*, the formula A the *free part* of P.

For example, $\underset{2\,4\,1}{\forall \exists \forall}\; \rho x^4 x^2 \vee (\sigma x^5 \wedge x^1 \equiv x^3)$ is prenex. In a prenex formula we distinguish between those of the integers $1, \ldots, n(\mathrm{A})$ which are indices of quantifiers, called the *indices of the prefix* U or *bound* indices, and the others, called the *free* indices. In the above example, $n(\mathrm{A}) = 5$, the indices 1, 2, 4 are bound, while 3 and 5 are free. The bound indices, in turn, are divided into $\forall$-*indices* and $\exists$-*indices* (in our example, 1 and 2 are $\forall$-indices and 4 is an $\exists$-index).

5.1.2. A formula is said to be *canonical* if, for each quantifier with indices $i_1, \ldots, i_r$ and the formula Q following it, the indices $i_1, \ldots, i_r$ are precisely the r last integers $\leqslant n(\mathrm{Q})$; in this case the arity of $\underset{i_1, \ldots, i_r}{\forall}$ Q or $\underset{i_1, \ldots, i_r}{\exists}$ Q is $n(\mathrm{Q}) - r$.

A formula is said to be *canonical prenex* if it is both prenex and canonical; thus the indices appear in increasing order in the usual notation, in decreasing order as to composition of quantifiers, and they assume consecutive values. Thus, if n is the arity of the formula and p the number of indices, the latter are $n+1, \ldots, n+p$. The integers $1, \ldots, n$ are then the *free indices,* $n+1, \ldots, n+p$ the *bound indices* of the canonical prenex formula.

5.2. Logical operator, value, representation by a prenex formula, active and inactive indices

With each logical formula P we associate the *operator represented by* P, which we denote by $\mathscr{P}$. Any $\mu(\mathrm{P})$-ary multirelation is said to be *assignable* to P, and transformed into an $n(\mathrm{P})$-ary relation, $\mathscr{P}(\mathrm{M})$, also denoted by $\mathrm{P}(\mathrm{M})$. The value of this relation for a sequence of elements $a_1, \ldots, a_n$ of the base of M is called the *value of* P *for* $(\mathrm{M}; a_1, \ldots, a_n)$, denoted by $\mathrm{P}(\mathrm{M})(a_1, \ldots, a_n)$.

When P reduces to a free operator $\mathscr{P}$, the transform $\mathrm{P}(\mathrm{M})$ is $\mathscr{P}(\mathrm{M})$.

When $\mathrm{P}=\alpha \mathrm{P}_1 \ldots \mathrm{P}_h$, where α is an h-ary connection and $\mathrm{P}_1, \ldots, \mathrm{P}_h$ the immediate subformulas of P, suppose we already have the transforms $\mathrm{P}_1(\mathrm{M}), \ldots, \mathrm{P}_h(\mathrm{M})$, which are relations of arities $n(\mathrm{P}_1), \ldots, n(\mathrm{P}_h)$; then the transform $\mathrm{P}(\mathrm{M})$ is

$$\alpha^{n(\mathrm{P})}_{n(\mathrm{P}_1), \ldots, n(\mathrm{P}_h)} (\mathrm{P}_1(\mathrm{M}), \ldots, \mathrm{P}_h(\mathrm{M})),$$

where $n(\mathrm{P})=$ arity of $\mathrm{P}=\max(n(\mathrm{P}_1), \ldots, n(\mathrm{P}_h))$. In other words, if we set

$$n_1=n(\mathrm{P}_1), \ldots, n_h=n(\mathrm{P}_h),$$

and

$$n=n(\mathrm{P})=\max(n_1, \ldots, n_h),$$

then the transform is the n-ary relation whose value for each n-tuple $(x_1, \ldots, x_n)$ is

$$\alpha(\mathrm{P}_1(\mathrm{M})(x_1, \ldots, x_{n_1}), \ldots, \mathrm{P}_h(\mathrm{M})(x_1, \ldots, x_{n_h})).$$

When $\mathrm{P}= \underset{i_1, \ldots, i_r}{\forall} \mathrm{Q}$ or $\underset{i_1, \ldots, i_r}{\exists} \mathrm{Q}$, where Q is the immediate subformula of P, suppose we already have the transform $\mathrm{Q}(\mathrm{M})$, which is of arity $n(\mathrm{Q})$; then the transform $\mathrm{P}(\mathrm{M})$ is $\underset{i_1, \ldots, i_r}{\forall^{n(\mathrm{P})}_{n(\mathrm{Q})}} \mathrm{Q}(\mathrm{M})$ or $\underset{i_1, \ldots, i_r}{\exists^{n(\mathrm{P})}_{n(\mathrm{Q})}} \mathrm{Q}(\mathrm{M})$, where $n(\mathrm{P})=$ arity of $\mathrm{P}=$ maximum integer $\leqslant n(\mathrm{Q})$ and distinct from $i_1, \ldots, i_r$.

In other words, the operator represented by P is obtained by composition in the sense of 3.5.2, associating with each occurrence of a free operator the operator itself, with each occurrence t of an h-ary connection α the connector defined by α, the arities of the h subformulas following t, and their maximum; finally, with each occurrence t of an indefinite quantifier q we associate the quantifier defined by q, the arity of the subformula following t, and the maximum integer not exceeding the arity and distinct from the indices of q.

A *logical operator* is an operator which is represented by some logical formula.

Examples. Free operators, quantifiers, the operator represented by $\underset{2}{\forall}((\neg \rho x^1 x^2) \bigvee x^1 \equiv x^2)$ which transforms any binary relation R into the unary relation: $S(x_1) = +$ if $R(x_1, x') = -$ for all $x' \neq x_1$; in particular, if R is an ordering, then $S(x_1) = +$ for maximal x_1.

Let M and M′ be two multirelations mapped onto each other by an isomorphism f; then, for any logical operator $\mathscr{P}$, the relations $\mathscr{P}(M)$ and $\mathscr{P}(M')$ are transformed into each other by f. To verify this, it suffices to note that this property is true of free operators and quantifiers. At the same time, an operator which commutes in this way with an isomorphism need not be logical; counterexample: the operator of transitive closure (see Volume 2, Section 1.7.3).

There are at most denumerable many logical operators of given arities, as this is the case for the free operators, connectors and quantifiers appearring in them.

5.2.1. *Let $\mathscr{P}$ be a (μ, m)-ary logical operator and f a mapping of $\{1, \ldots, m\}$ into $\{1, \ldots, n\}$; then the composite operator $f^n_m \mathscr{P}$ is logical.*

▷ If $\mathscr{P}$ is a free operator, the same holds for $f^n_m \mathscr{P}$ by 4.3.2.

If $\mathscr{P}$ has the form $\alpha^m_{m_1, \ldots, m_h}(\mathscr{P}_1, \ldots, \mathscr{P}_h)$, where $\mathscr{P}_i (1 \leqslant i \leqslant h)$ is of arity m_i and $m = \max(m_1, \ldots, m_h)$, let f_i denote the restriction of f to $\{1, \ldots, m_i\}$. By 3.5.4, we have

$$f^n_m \mathscr{P} = f^n_m \alpha^m_{m_1, \ldots, m_h}(\mathscr{P}_1, \ldots, \mathscr{P}_h) = \alpha^n_{n, \ldots, n}((f_1)^n_{m_1} \mathscr{P}_1, \ldots, (f_h)^n_{m_h} \mathscr{P}_h),$$

and this relation provides the necessary induction step.

If $\mathscr{P}$ is $\underset{i}{\forall}{}^m_m \mathscr{Q}$, where $i < m$, or $\underset{m+1}{\forall}{}^m_{m+1} \mathscr{Q}$, then, by 3.8.5, the composite operator $f^n_m \mathscr{P}$ is $\underset{n+1}{\forall}{}^n_{n+1} f'^{n+1}_m \mathscr{Q}$ or $\underset{n+1}{\forall}{}^n_{n+1} f''^{n+1}_{m+1} \mathscr{Q}$, and again it suffices to assume the statement true for $\mathscr{Q}$. ◁

5.2.2. *If $\mathscr{P}, \mathscr{Q}_1, \ldots, \mathscr{Q}_h$ are logical operators, then the composite operator $\mathscr{P}(\mathscr{Q}_1, \ldots, \mathscr{Q}_h)$ is a logical operator, provided it exists.*

▷ The statement is evident if $\mathscr{P}$ is a selector, + or − operator, identifier, or connector $\alpha^n_{m_1,\ldots,m_h}$ with $n = \max(m_1, \ldots, m_h)$.

By the preceding arguments, it is true if $\mathscr{P}$ is a rank-changer. Every free operator is obtained by composition of other operators (see 4.6.2), and so the statement is true if $\mathscr{P}$ is free. It is evident if $\mathscr{P}$ is a quantifier $\underset{i_1,\ldots,i_r}{\forall^n_m}$ or $\underset{i_1,\ldots,i_r}{\exists^n_m}$ with n = maximum integer $\leqslant m$ and distinct from the $i_1, \ldots, i_r$. Since every logical operator is obtained by composition of preceding operators, the statement is proved. ◁

5.2.3. *Let $\mathscr{P}$ be a (μ, m)-ary operator representable by a canonical prenex formula, and f a mapping of $\{1, \ldots, m\}$ into $\{1, \ldots, n\}$. Then the composite operator $f^n_m \mathscr{P}$ is representable by a canonical prenex formula.*

▷ If $\mathscr{P}$ is free, then so is $f^n_m \mathscr{P}$, and it is its own canonical prenex formula.

Otherwise, $\mathscr{P}$ has the form $\underset{m+1}{\forall^m_{m+1}} \mathscr{Q}$ or $\underset{m+1}{\exists^m_{m+1}} \mathscr{Q}$, where $\mathscr{Q}$ is a $(\mu, m+1)$-ary operator representable by a canonical prenex formula. By 3.8.5(2), there exists a mapping f'' of $\{1, \ldots, m+1\}$ into $\{1, \ldots, n+1\}$ such that

$$f^n_m \mathscr{P} = f^n_m \underset{m+1}{\forall^m_{m+1}} \mathscr{Q} = \underset{n+1}{\forall^n_{n+1}} f''^{n+1}_{m+1} \mathscr{Q}.$$

Proceeding by induction, assume the statement true for $\mathscr{Q}$, which is representable like $\mathscr{P}$ by a canonical prenex formula obtained from $\mathscr{P}$ by deleting a quantifier. Let Q be a canonical prenex formula representing $f''^{n+1}_{m+1} \mathscr{Q}$; then $\underset{n+1}{\forall}$ Q is canonical prenex and represents $f^n_m \mathscr{Q}$. The same holds with $\forall$ replaced by $\exists$. ◁

5.2.4. *Let $\mathscr{P}$ be a (μ, n)-ary operator and $\mathscr{P}'$ a (μ, n')-ary operator, $n' \leqslant n$, both representable by canonical prenex formulas. Then the operator $\bigwedge^n_{n,n'}(\mathscr{P}, \mathscr{P}')$ is representable by a canonical prenex formula.*

▷ Let UA denote a canonical prenex formula representing $\mathscr{P}$, in which the prefix U has indices $n+1, \ldots, n+p$ and the free part A has arity $n+p$. Let $d^{n+p}_{n'}$ be the dilator which transforms every n'-ary relation R into

$$\mathrm{S}(x_1, \ldots, x_{n'}, \ldots, x_{n+p}) = \mathrm{R}(x_1, \ldots, x_{n'}).$$

By 5.2.3, the operator $d_{n'}^{n+p}\,\mathscr{P}'$ is representable by a canonical prenex formula, U′A′, say, where U′ is the prefix, with indices beginning from $n+p+1$, and A′ is the free part.

Now UU′(A⋀A′) is a canonical prenex formula with arity n; it suffices to show that it represents $\bigwedge_{n,n'}^{n}(\mathscr{P}, \mathscr{P}')$. For this in turn it will suffice to show that it represents an operator equideducible from $\bigwedge_{n,n'}^{n}(\mathscr{P}, \mathscr{P}')$. The indices of U′ exceed the arity $n+p$ of A, hence they are inactive in any transform A(M) (where M is μ-ary). Thus we obtain an equideducible operator if we replace UU′(A⋀A′) by U(A⋀U′A′) (see 3.7.2). The indices of U run from $n+1 \geqslant n'+1$ to $n+p$; hence they are inactive in the operator $d_{n'}^{n+p}\,\mathscr{P}'$ represented by U′A′, or, more precisely, in any transform (U′A′)(M). Thus we obtain an equideducible formula if we replace U(A⋀U′A′) by (UA)⋀(U′A′). But this is the operator $\bigwedge_{n,n+p}^{n+p}(\mathscr{P}, d_{n'}^{n+p}\,\mathscr{P}')$, which is equideducible from $\bigwedge_{n,n'}^{n}(\mathscr{P}, \mathscr{P}')$. ◁

5.2.5. *Every logical operator is representable by a canonical prenex formula.*

In other words, it is the composite of a free operator with quantifiers.

This will follow from (1) through (4) below.

(1) Every free operator is a canonical prenex formula.

(2) If a (μ, n)-ary operator $\mathscr{P}$ is representable by a canonical prenex formula with prefix U and free part A, then $\neg_n^n \mathscr{P}$ is representable by the canonical prenex formula with prefix U′ obtained from U when ∀ and ∃ are interchanged, and free part A′ = ¬A.

(3) If $\mathscr{P}_1, \ldots, \mathscr{P}_h$ are operators of predicarity μ and arities $n_1, \ldots, n_h$, representable by canonical prenex formulas, and if α is an h-ary connection, then the operator $\alpha_{n_1,\ldots,n_h}^{n}(\mathscr{P}_1, \ldots, \mathscr{P}_h)$, with $n = \max(n_1, \ldots, n_h)$, is also so representable. This is seen by decomposing α into a superposition of ¬ and ⋀, and using (2) above and 5.2.4.

(4) If a (μ, n)-ary operator $\mathscr{P}$ is representable by a canonical prenex formula and $i < n$, then $\underset{i}{\forall}{}_n^n \mathscr{P}$ is so representable. In fact, by 3.8.5(1), with the identity for f, there exists a mapping f' such that

$$\underset{i}{\forall}{}_n^n \mathscr{P} = \underset{n+1}{\forall}{}_{n+1}^{n} f'^{\,n+1}_{\,n} \mathscr{P}.$$

By 5.2.3, the operator $f'^{\,n+1}_{\,n} \mathscr{P}$ is representable by a canonical prenex formula, and thus the same holds for this operator preceded by $\forall^n_{n+1}$. The same holds with $\forall$ replaced by $\exists$.

5.2.6. Let P be a formula, n its arity. An integer $i \leqslant n$ will be called an *active* or *inactive index* in P according to whether it is active or inactive in the operator represented by P. In other words, i is inactive in P if it is inactive in the relation P(M) for any multirelation M assignable to P (see 3.6.1). For convenience, we shall consider i to be inactive in P when $i > n$, in particular, when P is a subformula of a formula Q whose arity is $\geqslant i$.

5.3. Thesis, antithesis, explicit thesis, deduction, transform of a formula under a mapping

A (μ, n)-ary logical formula P is called a *thesis* if $\mathrm{P}(\mathrm{M})(x_1, \ldots, x_n) = +$ for any μ-ary multirelation M and any n-tuple $(x_1, \ldots, x_n)$ of base elements. We write $\vdash$P. An *antithesis* is defined similarly; the notation is P$\vdash$. When the base is empty, the only condition to be satisfied by a thesis is that it never assume the value $-$. In all cases, a (μ, n)-ary logical formula is a thesis if and only if its operator is equideducible from $+^0_\mu$.

If we start from a tautological connective formula and replace each atom at all its occurrences by a logical formula, we obtain a thesis which is also called a *tautology*; the same remark holds for a *contradiction*. We have already seen that $(x^1 \equiv x^2 \wedge \rho x^1) \Rightarrow \rho x^2$ is a thesis but not a tautology (see 4.6.3).

If P is a thesis, the same holds for $\mathscr{U}$P for any indefinite quantifier $\mathscr{U}$, with one or more indices. In particular, if A is a free thesis and $\mathscr{U}$ a quantifier, then $\mathscr{U}$A is a thesis. For example,

$$\underset{1}{\forall}\underset{2}{\exists}(\rho x^1 \vee \neg \rho x^1 \vee \rho x^2).$$

A necessary and sufficient condition for a formula P of arity n to be a thesis is that the bound formula $\underset{1,\ldots,n}{\forall}$ P be a thesis. Note that we must choose the quantifier $\forall$ that transforms the n-ary empty relation into $(\emptyset, +)$; see 3.6.2.

On the other hand, note that $\underset{1}{\exists} x^1 \equiv x^2$ is a thesis while $x^1 \equiv x^2$ is not.

Let P be a formula, Q a subformula of P (possibly with several occurrences in P), and x a free index of Q, also free in P, which may be active in Q but is inactive in all occurrences of subformulas of P disjoint from those of Q. For example, take $Q = \rho x \wedge x \equiv y$ and $P = (Q \vee y \equiv z) \vee \neg Q$. Suppose, moreover, that the occurrences of Q in P are dominated by connections (if $Q \neq P$) but not by any quantifier.

Under these assumptions, if P *is a thesis, the formula obtained by adding* $\underset{x}{\forall}$ *before each occurrence of* Q *is a thesis. The same holds with* $\underset{x}{\exists}$ *in place of* $\underset{x}{\forall}$.

▷ Let P* be the formula thus obtained, and suppose that it is not a thesis. Let M be a multirelation substituted for the predicates, and $b, c, \ldots$ elements of the base of M substituted for the free indices $y, z, \ldots$ of P other than x, such that $P^*(M)(b, c, \ldots) = -$. Now if the formula $\underset{x}{\forall} Q$ assumes the value + for $M, b, c, \ldots$, then Q assumes the value + for $M, a, b, c, \ldots$ where a is any element of the base. Thus the formulas $P^*(M)(b, c, \ldots)$ and $P(M)(a, b, c, \ldots)$ have the same truth value, because x is inactive at occurrences outside Q; thus the second expression becomes − like the first, so that P is not a thesis. On the other hand, if $\underset{x}{\forall} Q$ assumes the value − for $M, b, c, \ldots$, then there exists an element a of the base such that Q assumes the value − for $M, a, b, c, \ldots$. Thus the two expressions considered above again have the same truth value, since x is inactive at all occurrences outside Q and $y, z, \ldots$ remain free in P, so that $b, c, \ldots$ are again substituted for $y, z, \ldots$. Thus, again P is not a thesis. The same proof holds for $\exists$ in place of $\forall$. ◁

Note that the requirement that each occurrence of Q be associated with a quantifier $\underset{x}{\forall}$ is essential. Otherwise one could go, for example, from the thesis $\neg \rho x \vee \rho x$ with $Q = \rho x$ to the non-thesis $\neg \rho x \vee \underset{x}{\forall} \rho x$. It is also essential that no occurrence of Q be dominated by a quantifier; otherwise we could go from the thesis $\underset{y}{\exists} x \equiv y$, with $Q = (x \equiv y)$, to the non-thesis $\underset{y}{\exists} \underset{x}{\forall} x \equiv y$. The statement refers necessarily to occurrences of *one* formula Q and not only of formulas which are in some sense similar, say transforms of each other under a bijective change of indices; otherwise we could go from the thesis $x \not\equiv y \vee x \not\equiv z \vee y \equiv z$ to the non-thesis $(\underset{x}{\forall} x \not\equiv y) \vee (\underset{x}{\forall} x \not\equiv z) \vee y \equiv z$.

We define a *quasi-free* thesis to be a formula obtained from a free thesis by a finite number of adjunctions of quantifiers as described in the preceding proposition. It is clear that every tautology is a quasi-free thesis (it is obtained from a connective tautology, in the sense of 2.2, by substituting a logical formula for each atom). Similarly, we see that every free thesis preceded by quantifiers is quasi-free. Apart from these two important cases, there are other quasi-free theses; for example:

$$\underset{x}{\forall}\rho x \bigvee \left(\neg \underset{x}{\forall}\rho x \bigwedge y \equiv z\right) \bigvee y \not\equiv t \bigvee z \not\equiv t.$$

One can decide in finitely many steps whether a formula is a quasi-free thesis: when the quantifiers are omitted, the result should be a free thesis, and this can be decided in finitely many steps (4.6.3); one then restores the quantifiers one by one in an order in which all the conditions of the preceding proposition are satisfied for each.

5.3.1. A prenex formula UA, with prefix U and free part A, is called an *explicit thesis* if there exists a function f which associates with each $\exists$-index i an integer $f(i)$ which is either the index of an earlier quantifier in the sequence U, or a free index; $f(j)=j$ for any other index ($\forall$-index or free index); and the transform $\mathrm{A}]f$ of the free part A under f is a free thesis.

In the case of a canonical prenex formula, the conditions on f are: $f(i) \leqslant i$ for each $\exists$-index i, and $f(j)=j$ for any other index. To see that UA is a thesis, we note that $\mathrm{UA}]f$ is a thesis (see above) and that $\mathrm{UA}]f \vdash \mathrm{UA}$ by 3.8.6. Indeed, by 4.6.5, the operator represented by $\mathrm{A}]f$ is $f^n_n\mathscr{A}$, where n is the arity of A and $\mathscr{A}$ is represented by A.

Examples. $\underset{1\,2}{\forall\exists} x^1 \equiv x^2$ is an explicit thesis: taking $f(2)=1$, we find that $\mathrm{A}]f$ is the free thesis $x^1 \equiv x^1$, i.e., $+(x^1)$. Similarly, $\underset{1\,2}{\forall\exists}\rho x^1 \bigvee \neg \rho x^2$ leads to the free thesis $\rho x^1 \bigvee \neg \rho x^1$.

5.3.2. A thesis need be neither a tautology nor an explicit thesis. An example is $\underset{1\,2}{\exists\forall}\rho x^1 \bigvee \neg \rho x^2$; another example:

$$\underset{1\,2\,3}{\exists\forall\forall} \neg \rho x^1 \bigvee (\rho x^2 \bigwedge \rho x^3),$$

or a more complicated thesis:

$$\underset{1\,2\,3\,4}{\exists\forall\exists\exists} \neg \rho x^1 x^2 \bigvee (\rho x^2 x^3 \bigwedge \rho x^3 x^4).$$

Interpretation: given a relation R, either there exists x_1 such that $R(x_1, x_2) = -$ for all x_2, or for all x_2 there exist x_3 and x_4 such that $R(x_2, x_3) = R(x_3, x_4) = +$.

Some other examples will be given in Exercise 4.

A logical formula which is not an antithesis is said to be *consistent*. We have already seen that ρx^1, for example, is consistent, though not a thesis (see 4.6.3).

One can decide in finitely many steps whether a connective formula is or is not a tautology (assign all possible sequences of + and − to the atoms). Similarly, one can decide if a free formula is or is not a thesis (see 4.6.3). It follows that one can decide whether a logical prenex formula is or is not an explicit thesis. To do this, write the formula in canonical form UA (prefix U, free part A); consider all the functions f (there are only finitely many!) defined on the indices of A, such that $f(i) \leqslant i$ for every $\exists$-index i and $f(i) = i$ for every $\forall$-index or free index; then check whether there exists a mapping f such that $A]f$ is a free thesis.

On the other hand, we know of no algorithm by which one can decide whether a logical formula is or is not a thesis. We even have reason, well grounded in experience, to believe that no such algorithm exists.

5.3.3. If P and Q are logical formulas of the same predicarity and arbitrary arities, we shall say, as in 4.6.4, that Q is *deducible from* P, writing $P \vdash Q$, if $\mathscr{P}_P \vdash \mathscr{P}_Q$, where $\mathscr{P}_P$ and $\mathscr{P}_Q$ are the operators represented by P and Q.

The remarks in 4.6.4 that the deducibility relation $\vdash$ is a pre-ordering and the equideducibility relation $-\vdash$ is an equivalence relation are also valid for logical formulas. In particular, deducibility $P \vdash Q$ reduces to the condition that $P \Rightarrow Q$ is a thesis, and hence, if we denote the larger of the arities of P and Q by n, to the condition that the bound formula $\forall_{1, \ldots, n}\ P \Rightarrow Q$ be a thesis.

If $P \vdash Q$ and $\mathscr{U}$ is an indefinite quantifier with one or more indices, then $\mathscr{U}P \vdash \mathscr{U}Q$ (see 3.6.3).

Let P be a logical formula, the only connections in which are $\neg, \bigwedge, \bigvee$. Let P′ be the formula obtained from P by interchanging $\bigwedge$ and $\bigvee$, $\forall$ and $\exists$, and adding $\neg$ before each free operator; then P′ is equideducible from $\neg$P and has the same arities.

For every logical formula there is at least one formula containing only the connections $\neg, \bigwedge$ (or only $\neg, \bigvee$) and representing the same operator. Thus, if a class contains every free operator, is closed under the operations $\neg, \bigwedge, \underset{i}{\forall}$ for all integers i, and also under equideducibility, then it contains every logical formula.

5.3.4. Let P be a logical formula of arity m and $\bar{m} \geqslant m$ the maximum arity of its free operators. Let the indices of the quantifiers be $\leqslant \bar{m}$ (otherwise, the quantifier reduces to the identity operator). Let f be a mapping defined on $\{1, \ldots, \bar{m}\}$. Replace each free operator $\mathscr{A}$, of arity $p \leqslant \bar{m}$, by its transform $f_p^{p'} \mathscr{A}$, where $p' = \max(f(1), \ldots, f(p))$, and each index $j \leqslant \bar{m}$ of each quantifier by $f(j)$. The resulting formula will be the *transform of* P *under* f, denoted by P$]f$.

If f is injective on $\{1, \ldots, \bar{m}\}$, the operator represented by P$]f$ *is equideducible from the composite operator $f_m^{m'} \mathscr{P}$, where $\mathscr{P}$ is the operator represented by* P, *m is the arity of* P, *and $m' = \max(f(1), \ldots, f(m))$.*

▷ If P is free, this reduces to 4.6.5.

If P has the form $\alpha \mathrm{P}_1 \ldots \mathrm{P}_h$, with α an h-ary connection, let $\mathscr{P}_i$ $(1 \leqslant i \leqslant h)$ denote the operator represented by P_i, $\mathscr{P}'$ and $\mathscr{P}'_i$ the operators represented by the transforms P$]f$ and $\mathrm{P}_i]f$, m_i the arity of P_i, m'_i the maximum of $f(1), \ldots, f(m_i)$, m the arity of P $(= \max(m_1, \ldots, m_h))$ and m' the maximum of $f(1), \ldots, f(m)$ $(= \max(m'_1, \ldots, m'_h))$. By assumption, $\mathscr{P}'_i -\vdash f_{m_i}^{m'_i} \mathscr{P}_i$, and so

$$\mathscr{P}' -\vdash \alpha_{m'_1, \ldots, m'_h}^{m'} (f_{m_1}^{m'_1} \mathscr{P}_1, \ldots, f_{m_h}^{m'_h} \mathscr{P}_h) = \alpha_{m', \ldots, m'}^{m'} (f_{m_1}^{m'} \mathscr{P}_1, \ldots, f_{m_h}^{m'} \mathscr{P}_h).$$

Using 3.5.4, we get

$$\mathscr{P}' -\vdash f_m^{m'} \alpha_{m_1, \ldots, m_h}^{m} (\mathscr{P}_1, \ldots, \mathscr{P}_h) = f_m^{m'} \mathscr{P}.$$

If P has the form $\underset{i}{\forall}$Q, let m be the arity of Q and m' the maximum of $f(1), \ldots, f(m)$. Then, by assumption, if $\mathscr{Q}$ is the operator represented by Q,

$$\mathscr{P}' -\vdash \underset{f(i)}{\forall}{}_{m'}^{m'} f_m^{m'} \mathscr{Q} = f_m^{m'} \underset{i}{\forall}{}_m^m \mathscr{Q}$$

by 3.8.1, for f being injective, takes the value $f(i)$ only once; and moreover $\underset{i}{\forall}{}_m^m \mathscr{Q} -\vdash \mathscr{P}$.

Now, either $i < m$, in which case the arity of P is m and $\underset{i}{\forall}{}_m^m \mathscr{Q} = \mathscr{P}$, so that $\mathscr{P}' -\vdash f_m^{m'} \mathscr{P}$; or $i = m$, and then the arity of P is $m-1$. Letting

d^m_{m-1} denote the $(m-1, m)$-ary dilator, we have

$$\underset{m}{\forall}{}^m_m \mathcal{Q} = d^m_{m-1}\mathcal{P}, \quad \text{and thus} \quad \mathcal{P}' - \vdash f^{m'}_m d^m_{m-1}\mathcal{P}.$$

Therefore, setting

$$m'' = \max(f(1), \ldots, f(m-1)) \leqslant m',$$

we get

$$f^{m'}_m d^m_{m-1} = f^{m'}_{m-1} - \vdash f^{m''}_{m-1}, \quad \text{and thus} \quad \mathcal{P}' - \vdash f^{m''}_{m-1}\mathcal{P}.$$

The same holds if P has the form $\underset{i}{\exists}$Q. ◁

Let P *be a thesis, m* the maximum arity of its free operators, and *f* an *injective mapping* defined on $\{1, \ldots, m\}$, and assume that the indices of the quantifiers are $\leqslant m$. Then, by the preceding arguments, *the transform of* P *under f is a thesis.* For example, starting from $\underset{1}{\forall}\underset{2}{\exists} x^1 \equiv x^2$ and taking $f(1)=3$, $f(2)=2$, we get the thesis $\underset{3}{\forall}\,\underset{2}{\exists} x^2 \equiv x^3$.

Note that if f is not injective, the preceding result is no longer valid (in contrast to the situation for a free formula: see 4.6.5). For example, $(\underset{2}{\exists} x^1 \not\equiv x^2) \bigvee \underset{3,4}{\forall} x^3 \equiv x^4$ is a thesis; applying the mapping defined by $f(1)=2$ and $f(i)=i$ for $i=2, 3, 4$, we get $(\underset{2}{\exists} - {}^2) \bigvee \underset{3,4}{\forall} x^3 \equiv x^4$, which is no longer a thesis.

5.4. Model of a Bound Formula, Logical Class

We now consider the case in which the logical formula P is bound; thus, for every multirelation M assignable to P, the relation P(M) is 0-ary: it reduces either to (E, +) or to (E, −), where E = base of M. In other words, with each M we associate a value + or −, called the *value of* P *for* M and denoted by P(M). Instead of saying that P(M) = +, we shall also say that M *satisfies* P or that M is a *model* of P.

If M is a model of P, then every isomorphic image of M is also a model of P.

The class of μ-ary multirelations which give P the value + is called a *logical class*, denoted by $\mathscr{A}_P$. We shall say that P *represents* the class $\mathscr{A}_P$.

The union, intersection of two logical classes, the complement of a logical class (relative to the class of all multirelations of the given arity) are logical classes.

If M has a finite base, then, together with all its isomorphic images, it constitutes a logical class. In other words, there exists a bound formula whose only models are M and its isomorphic images. To see this, let p denote the (finite) cardinality of M, and consider the formulas $x^i \not\equiv x^j$, where $i, j = 1, \ldots, p$ and $i < j$. Form the formula

$$\underset{p+1}{\forall}\, x^1 \equiv x^{p+1} \bigvee \cdots \bigvee x^p \equiv x^{p+1}.$$

Then, for each m-ary relation R derived from M and each m-tuple $i_1, \ldots, i_m$ of numbers from 1 to p, take the formula $\rho x^{i_1} \ldots x^{i_m}$ itself or preceded by $\neg$, according to whether $R(i_1, \ldots, i_m) = +$ or $-$; finally, consider the conjunction of all these formulas, preceded by $\underset{1,\ldots,p}{\exists}$.

It follows that every finite set of multirelations of given arity and finite bases, together with their isomorphic images, constitute a logical class.

The logical classes of given arity constitute a denumerable set, since the set of logical formulas is denumerable and, for each integer p, for example, we have a logical class consisting of multirelations of cardinal p.

Examples. The universal classes of 4.8; the class of reflexive binary relations, represented by $\underset{1}{\forall} \rho x^1 x^1$; the class of transitive binary relations; the class of symmetric relations, represented by $\underset{1,2}{\forall}\, \rho x^1 x^2 \Rightarrow \rho x^2 x^1$.

The class of antisymmetric relations; the class of totally connected relations, represented by $\underset{1,2}{\forall}\, \rho x^1 x^2 \bigvee \rho x^2 x^1$.

The class of pre-orderings (reflexive and transitive relations), the class of equivalence relations (or symmetric pre-orderings), the class of orderings (antisymmetric pre-orderings), the class of chains or total orderings.

The class of functional binary relations, i.e., all R such that for all x_1 in the base there is a unique x_2 such that $R(x_1, x_2) = +$: take the conjunction of the formulas

$$\underset{1}{\forall}\underset{2}{\exists} \rho x^1 x^2 \quad \text{and} \quad \underset{1,2,3}{\forall} (\rho x^1 x^2 \bigwedge \rho x^1 x^3) \Rightarrow x^2 \equiv x^3.$$

If $P \vdash Q$, where P and Q are bound formulas, then $\mathscr{A}_P \subseteq \mathscr{A}_Q$, and conversely. For example, the formula $Q = \underset{1}{\forall} \rho x^1 x^1$ is deducible from the conjunction of

$$\underset{1,2,3}{\forall}(\rho x^1x^2\wedge\rho x^2x^3)\Rightarrow\rho x^1x^3, \qquad \underset{1,2}{\forall}\rho x^1x^2\Rightarrow\rho x^2x^1,$$

and

$$\underset{1}{\forall}\underset{2}{\exists}\rho x^1x^2.$$

An equivalent formulation of this statement is that the class of symmetric transitive relations with the property: for each x_1 there is at least one x_2 such that $R(x_1, x_2)=+$, is a subclass of the class of reflexive relations. Indeed, it is the class of equivalence relations.

5.4.1. *The class of birelations* (R, S) *in which* R *and* S *are two relations of common arity and are identical, is a logical class.*

Indeed, if m is the arity of R and S, let ρ and σ denote two m-ary predicates. It suffices to consider the formula $\underset{1\ldots m}{\forall}\rho x^1\ldots x^m\Leftrightarrow\sigma x^1\ldots x^m$.

The statement immediately generalizes to the case of multirelations of given arity in which relations of certain ranks are identical.

5.4.2. If $\mathscr{A}$ is a logical class and $\mathscr{P}_1,\ldots,\mathscr{P}_h$ is a finite sequence of logical operators, then the *inverse image of* $\mathscr{A}$ *under* $\mathscr{P}_1,\ldots,\mathscr{P}_h$ *is a logical class* (the inverse image is defined in 3.5.6). Indeed, $\mathscr{A}$ can be regarded as a 0-ary logical operator, writing for every multirelation N of the given arity $\mathscr{A}(\mathrm{N})=+$ or $-$ instead of $\mathrm{N}\in\mathscr{A}$ or $\mathrm{N}\in\mathscr{A}$. The inverse image then coincides with the composite operator $\mathscr{A}(\mathscr{P}_1,\ldots,\mathscr{P}_h)$, which is logical by 5.2.2.

Let $\mathscr{P}$ *be a logical operator; then the class of birelations* (R, $\mathscr{P}$(R)) *is a logical class.* Indeed, it is the inverse image, under $\mathscr{P}$ and the identity, of the logical class of birelations (S, S), where S has the same arity as $\mathscr{P}$ (see 3.5.6 and 5.4.1).

5.4.3. Let $\mathscr{A}$ be a logical class of μ-ary multirelations and $\mathscr{P}$ an operator of predicarity μ. *Then the image of* $\mathscr{A}$ *under* $\mathscr{P}$ (*defined in* 3.5.6) *is not necessarily a logical class, even if* $\mathscr{P}$ *is free.*

For example, consider the class of chains without maximal element. This is a logical class, as shown by taking the conjunction of the formulas given above, which define the class of chains, and the formula

$$\underset{1\,2}{\forall\exists} x^1 \not\equiv x^2 \wedge \rho x^1 x^2 .$$

Take the free operator +, which associates with each binary relation R the 0-ary relation taking the value + on the base of R. Since chains without maximal element have infinite bases, if we admit the axiom of choice (or even only an axiom ensuring the existence of a chain over any set), this leads to the class of infinite sets, which is not a logical class (see 5.5.2 below).

Another example, concerning the chain of natural numbers and the successor relation, will be considered in Volume 2, Section 1.6.4.

Remark. If $\mathscr{P}$ is a logical operator, and $\mathscr{A}$ the class of all multirelations assignable to $\mathscr{P}$, then $\mathscr{P}(\mathscr{A})$ is not always a logical class (see 4.8.8).

5.5. Logical class: case of sets and unary multirelations

5.5.1. (1) *Let* P *be a prenex formula of empty predicarity* (i.e., the multirelations assignable to P are simply sets). *Let* n *be the arity of* P, $\bar{n} \geqslant n$ *the maximum arity of the free operators of* P. *Then, for any two sets* E, E′ *containing at least* $\bar{n}$ *elements and two sequences,* $a_1, \ldots, a_n$ *from* E *and* $a'_1, \ldots, a'_n$ *from* E′, *where the second is the image of the first under a bijective mapping, we have*

$$\mathrm{P(E)}\,(a_1, \ldots, a_n) = \mathrm{P(E')}\,(a'_1, \ldots, a'_n).$$

(2) *In particular, let* P *be a bound prenex formula and* p *the maximum arity of its free operators. Then, for any two sets* E, E′ *with at least* p *elements,* P(E) = P(E′).

▷ We may always assume that P is canonical prenex, the indices of the quantifiers being $n+1, \ldots, n+p = \bar{n}$. If $p=0$, then P is free and the bijective mapping of a_i onto $a'_i (i=1, \ldots, n)$ is a local isomorphism of E onto E′, therefore of P(E) onto P(E′), which implies the desired equality.

Now assume the statement true for $n+1$ and p, and let us show that it is true for n and $p+1$. Let Q be the immediate subformula of P, say, $\mathrm{P} = \underset{n+1}{\forall}\,\mathrm{Q}$, where Q is of arity $n+1$ and its prefix has p indices. Suppose that $\mathrm{P(E)}\,(a_1, \ldots, a_n) = +$; then

$$\mathrm{Q(E)}\,(a_1, \ldots, a_n, x) = + \quad \text{for any} \quad x \in E.$$

We claim that then

$$Q(E')(a'_1, \ldots, a'_n, x') = + \quad \text{for any} \quad x' \in E'.$$

Indeed, E and E′ contain at least $\bar{n} = n + p + 1$ elements, and so there exists $y \in E$ such that the mapping taking each a_i onto a'_i $(i = 1, \ldots, n)$ and y onto x' is bijective. Hence, by the induction hypothesis (for $n+1$ and p),

$$Q(E')(a'_1, \ldots, a'_n, x') = Q(E)(a_1, \ldots, a_n, y) = +.$$

Consequently,

$$P(E')(a'_1, \ldots, a'_n) = +, \quad \text{therefore} \quad = P(E)(a_1, \ldots, a_n).$$

The same reasoning holds if we replace + by − and the phrase "for any x (or x')" by "for some x (or x')". Finally, the same reasoning is valid for $P = \underset{n+1}{\exists} Q$. ◁

5.5.2. Let $\mathscr{A}$ be a class of sets; then *$\mathscr{A}$ is a logical class only if one of the following two conditions holds:*

(1) There exists a finite set of integers $n_1, \ldots, n_h$ such that *$\mathscr{A}$ consists entirely of finite sets of cardinality $n_1, \ldots, n_h$.*

(2) There exist a finite set of integers $n_1, \ldots, n_h$ and an integer p greater than $n_1, \ldots, n_h$ such that *$\mathscr{A}$ consists of finite sets of cardinality $n_1, \ldots, n_h$ and of all sets of cardinality $\geqslant p$* (so that $\mathscr{A}$ contains all infinite sets).

▷ The classes of (1) are logical by 5.4 (this is the special case in which the multirelations M coincide with their bases). The classes of (2) are unions of a class of type (1) and the class of sets of cardinality $\geqslant p$, which can be represented by the formula

$$\underset{1,\ldots,p}{\exists} x^1 \not\equiv x^2 \wedge \cdots \wedge x^1 \not\equiv x^p \wedge x^2 \not\equiv x^3 \wedge \cdots \wedge x^{p-1} \not\equiv x^p.$$

Hence these are logical classes.

Conversely, for any logical class $\mathscr{A}$ consisting of sets, there exists a bound prenex formula P of empty predicarity which represents $\mathscr{A}$. Let p be the number of indices of the quantifiers, equal to the maximum arity of the free operators. By 5.5.1 (2), if E and E′ are two sets containing at least p elements, then $P(E) = P(E')$. Hence, from cardinality p on, either all sets or none are in $\mathscr{A}$. In view of the fact that two sets of the same cardinality assign the same value to P, this means that either case (1) or (2) in the statement of our theorem must hold. ◁

5.5.3. A multirelation $M=(R_1,\ldots, R_h)$ is said to be *unary* if all the component relations $R_i (i=1,\ldots, h)$ are unary (i.e., of arity 1). The base E of M is thus partitioned into at most 2^h equivalence classes, on each of which all relations R_i assume the same value (+ or −). These will be called the *classes associated* with M. Given another multirelation $M'=(R'_1,\ldots, R'_h)$ of the same arity, with base E′, a class associated with M and a class associated with M′ are said to *correspond* if the relations R_i and R'_i with the same index i assume the same value on both classes.

(1) *Let* P *be a prenex formula with h unary predicates, of arity n; let* $\bar{n}$ *be the maximum arity of its free operators. Consider two unary multirelations* $M=(R_1,\ldots, R_h)$ *with base* E *and* $M'=(R'_1,\ldots, R'_h)$ *with base* E′ *such that every two corresponding classes either have the same cardinality* $<\bar{n}$ *or cardinalities* $\geqslant\bar{n}$. *Let* $a_1,\ldots, a_n$ *and* $a'_1,\ldots, a'_n$ *be sequences of elements of* E *and* E′, *respectively, such that the mapping taking each* a_i *onto* a'_i $(i=1,\ldots,n)$ *is bijective, and for each i the elements* a_i *and* a'_i *are in corresponding classes. Then*

$$P(M)(a_1,\ldots, a_n)=P(M')(a'_1,\ldots, a'_n).$$

(2) *In particular, let* P *be a bound prenex formula with h unary predicates, p the maximum arity of its free operators. Let* $M=(R_1,\ldots, R_h)$ *and* $M'=(R'_1,\ldots, R'_h)$ *be two unary multirelations such that every two corresponding classes have the same cardinality* $<p$ *or cardinalities* $\geqslant p$. *Then* $P(M)=P(M')$.

The proof may be modeled on that of 5.5.1, with the following modifications. If $p=\bar{n}-n=0$, then P is free, the bijective mapping of a_i onto a'_i is a local isomorphism of M onto M′, hence of P(M) onto P(M′). One then shows that if the statement is true for $n+1$ and p, it is true for n and $p+1$. Replace E by M and E′ by M′ in P(E), P(E′), Q(E), Q(E′). To show that $Q(M')(a'_1,\ldots, a'_n, x')=+$, note that there exists y in the class corresponding to that of x' such that the mapping taking each a_i onto a'_i $(i=1,\ldots, n)$ and y onto x' is bijective, and takes the elements of a class onto elements of the corresponding class. It then follows from the hypothesis concerning $n+1$ and p that

$$Q(M')(a'_1,\ldots, a'_n, x')=Q(M)(a_1,\ldots, a_n, y).$$

5.5.4. *Let* P *be a bound prenex formula with unary predicates, h the number of relations in* P *and p the maximum arity of its free operators. Then* P *is a thesis if and only if* $\mathrm{P(M)}=+$ *for all multirelations whose base consists of at most $p.2^h$ elements*

In fact, with each multirelation M having h relations we associate a restriction M′ of M defined as follows: retain the classes associated with M which have $<p$ elements, and replace those which have $\geqslant p$ elements by a subset having exactly p elements. Then, by 5.5.3 (2), we have $\mathrm{P(M')}=\mathrm{P(M)}$. Thus the condition of the theorem implies that $\mathrm{P(M)}=+$ for every M.

This implies the following algorithm for deciding if a given bound unary formula is a thesis or not. Replace the formula by an equivalent prenex formula. Let h be the number of predicates and p the maximum arity of the free operators. There are only finitely many multirelations defined on the set $\{1, 2,\ldots, p.2^h\}$ or its subsets, and so one can check whether the formula assumes the value + for each of them.

5.5.5. *There exists an operator $\mathscr{P}$ which is logical, while $\not\mathscr{F}\mathscr{P}$ is not logical* (see 4.5 for the definition of $\not\mathscr{F}$, and compare 4.5.4 – the analog for a free operator). Indeed, consider the logical operator $\mathscr{P}$ which converts each binary relation R into the 0-ary relation with the same base such that $\mathscr{P}(\mathrm{R})=+$ if R is not a chain or if R is a chain with a maximal element, $\mathscr{P}(\mathrm{R})=-$ otherwise. The operator $\not\mathscr{F}\mathscr{P}$ assigns each set E the value + if E is finite (since then every chain over E has a maximal element) and the value – if E is infinite. By 5.5.2, we know that this is not a logical operator.

We shall prove in Volume 2, Section 1.7.3, that the operator of transitive closure is not logical. This is the operator which converts each binary relation R into the relation S defined thus: $\mathrm{S}(x, y)=+$ if there exists a finite sequence of elements $x_1=x, x_2,\ldots, x_r=y$ such that $\mathrm{R}(x_i, x_{i+1})=+$ for $i=1, 2,\ldots, r-1$.

5.6. Logical equivalence; denumerable-model theorem ([LOW, 1915; SKO, 1920])

Two multirelations M and M′ of the same arity μ are said to be *logically equivalent* if they assign the same value, + or –, to any bound formula of predicarity μ. This relation is obviously reflexive, symmetric, and transitive.

In view of the foregoing sections, the logical equivalence classes for sets (multirelations of empty arity) are the following: for each finite integer p, the class of sets of cardinality p, and in addition the class of all infinite sets. The former are special logical classes, while the latter is not a logical class. Two unary multirelations are logically equivalent if and only if every two of their corresponding classes (see 5.5.3) are of the same finite cardinality, or of infinite (not necessarily equal) cardinalities. For multirelations over finite bases, logical equivalence coincides with isomorphism. Similarly, by the foregoing sections this is also true for unary multirelations over denumerable bases.

On the other hand, two denumerable binary relations may be logically equivalent but not isomorphic. For example: the chain of natural numbers, and the same chain followed by a chain isomorphic to that of the integers. In general: all chains containing a minimal element and no maximal element, which are also discrete (every element has a successor and every element except the minimal one has a predecessor), are logically equivalent (see Volume 2, Sections 1.2.2 and 1.4; the result is due to [TAR, 1934]). Another example: The sums of rational numbers, algebraic numbers, and real numbers are three logically equivalent ternary relations; the birelations (sum, product) over the algebraic numbers and over the real numbers are logically equivalent [TAR, 1951].

Every μ-ary logical equivalence class is defined by a partition of the set of μ-ary formulas into two subsets (each formula of a subset assumes the value + for all the multirelations of the class, or the value − for all of them). It follows that for a given arity these classes constitute a set of cardinality at most that of the continuum. The set actually has the cardinality of the continuum, provided the arity μ contains an integer $\geqslant 2$. Let us prove this for $\mu=(2, 1)$, i.e., for birelations (R, S), where R is binary and S unary. Let R be the successor relation on the natural numbers and S an arbitrary unary relation on the same set. It suffices to note that, if S′ is a unary relation distinct from S, then (R, S) and (R, S′) cannot be logically equivalent. Take the smallest integer p such that $S'(p) \neq S(p)$, say $S(p) = +$ and $S'(p) = -$; then the following formula assumes the value + for (R, S), and the value − for (R, S′):

$$\underset{0,1,\ldots,p}{\exists}\left(\left(\underset{p+1}{\forall}\neg\rho x^{p+1}x^{0}\right)\wedge\rho x^{0}x^{1}\wedge\rho x^{1}x^{2}\wedge\cdots\wedge\rho x^{p-1}x^{p}\wedge\sigma x^{p}\right)$$

(for convenience, we have numbered the indices from 0 instead of from 1). This example again shows that there exist logical equivalence classes which are not logical classes, since the set of the latter is only denumerable.

5.6.1. Consider a canonical prenex formula with prefix U and free part A, a multirelation M with base E, and a sequence of elements $a_1,\ldots,a_n$ of E (where n is the arity of UA). Let p be the number of indices of U, and let $\mathscr{C}(\mathrm{U}, \mathrm{A}, \mathrm{M}, a_1,\ldots,a_n)$ denote the condition given by the conjunction of the following conditions $\mathscr{C}_1$ and $\mathscr{C}_2$:

($\mathscr{C}_1$) For each $\exists$-index i of U, there exists a function f_i which associates an element of E with every sequence of elements of E with indices $n+1$, $n+2,\ldots,i-1$.

($\mathscr{C}_2$) We have $\mathrm{A}(\mathrm{M})(a_1,\ldots,a_n,b_{n+1},\ldots,b_{n+p})=+$ for all $b_{n+1},\ldots,b_{n+p}$ in E such that $b_i=f_i(b_{n+1},\ldots,b_{i-1})$.

(1) *If* $\mathscr{C}(\mathrm{U}, \mathrm{A}, \mathrm{M}, a_1,\ldots,a_n)$ (i.e., $\mathscr{C}_1$ *and* $\mathscr{C}_2$) *is satisfied, then* $\mathrm{UA}(\mathrm{M})(a_1,\ldots,a_n)=+$.

(2) *If we admit the axiom of choice and* $\mathrm{UA}(\mathrm{M})(a_1,\ldots,a_n)=+$, *then* $\mathscr{C}(\mathrm{U}, \mathrm{A}, \mathrm{M}, a_1,\ldots,a_n)$ *is satisfied; thus the two conditions considered here are equivalent.*

Proof of (1). If U is empty ($p=0$), the conclusion reduces to the equality $\mathrm{A}(\mathrm{M})(a_1,\ldots,a_n)=+$, which is exactly the assumption $\mathscr{C}_2$. Suppose now that $\mathscr{C}$ implies $\mathrm{UA}(\mathrm{M})(a_1,\ldots,a_n)=+$ for a given prefix U with p indices, and let us show that the same holds for U preceded by $\underset{n}{\forall}$ or $\underset{n}{\exists}$, hence for $p+1$ indices.

We first assume that $\mathscr{C}(\underset{n}{\forall}\mathrm{U}, \mathrm{A}, \mathrm{M}, a_1,\ldots,a_{n-1})$ holds; note that then $\mathscr{C}(\mathrm{U}, \mathrm{A}, \mathrm{M}, a_1,\ldots,a_{n-1},a)$ for every a in E. In fact, for every $\exists$-index i of U there exists a function f_i' which is the previous function $f_i(a,\ldots)$ with the value a at the n-th position. This gives ($\mathscr{C}_1$) with indices $n+1,\ldots,i-1$. It also gives the equality ($\mathscr{C}_2$), i.e., $\mathrm{A}(\mathrm{M})(a_1,\ldots,a_{n-1},a,b_{n+1},\ldots,b_{n+p})=+$ for all $b_{n+1},\ldots,b_{n+p}$ in E such that

$$b_i=f_i(a,b_{n+1},\ldots,b_{n+p})=f_i'(b_{n+1},\ldots,b_{n+p}).$$

Since the assertion is assumed true for U, we have

$$\mathrm{UA}(\mathrm{M})(a_1,\ldots,a_{n-1},a)=+$$

for every a in E, and so $\underset{n}{\forall}\mathrm{UA(M)}\,(a_1,\ldots,a_{n-1})=+$.

Now assume that $\mathscr{C}(\underset{n}{\exists}\,\mathrm{U, A, M}, a_1,\ldots,a_{n-1})$ holds; note that then there exists a in E such that $\mathscr{C}(\mathrm{U, A, M}, a_1,\ldots,a_{n-1},a)$. By assumption, for the index n there exists a function over empty sequences assuming the sole value a in E. Moreover, for each $\exists$-index i of U $(n+1\leqslant i\leqslant n+p)$, there exists a function f_i' which is the previous function $f_i(a,\ldots)$. This gives $(\mathscr{C}_1)$ and $(\mathscr{C}_2)$ as before. Since the assertion is true for U, we have $\mathrm{UA(M)}\,(a_1,\ldots,a_{n-1},a)=+$ for some a in E, and so

$$\underset{n}{\exists}\mathrm{UA(M)}\,(a_1,\ldots,a_{n-1})=+.$$

Proof of (2). If U is empty $(p=0)$, the condition $\mathscr{C}$ reduces to $\mathrm{A(M)}\,(a_1,\ldots,a_n)=+$, i.e., to the assumption of the theorem. Suppose that $\mathrm{UA(M)}\,(a_1,\ldots,a_n)=+$ implies $\mathscr{C}$ for a given prefix U with p indices, and let us show that the same holds for U preceded by $\underset{n}{\forall}$ or $\underset{n}{\exists}$, assuming the axiom of choice.

We first assume that $\underset{n}{\forall}\mathrm{UA(M)}\,(a_1,\ldots,a_{n-1})=+$. It follows that, for every a in E, we have $\mathrm{UA(M)}\,(a_1,\ldots,a_{n-1},a)=+$, and hence by assumption the condition $\mathscr{C}(\mathrm{U, A, M}, a_1,\ldots,a_{n-1},a)$ holds. Hence, for every a and every $\exists$-index i of U, there exist functions $f_{a,i}$ satisfying $\mathscr{C}_1$ and $\mathscr{C}_2$. We define a function g_i on the indices $n, n+1,\ldots,i-1$, setting $g_i(a,\ldots)$ equal to one of the functions $f_{a,i}(\ldots)$ for each a (axiom of choice!). Then the condition $(\mathscr{C}_1)$ holds for the indices $n, n+1,\ldots,i-1$ and $(\mathscr{C}_2)$ is satisfied with

$$\mathrm{A(M)}\,(a_1,\ldots,a_{n-1},b_n,b_{n-1},\ldots,b_{n+p})=+$$

for every $b_n=a$ and all $b_{n+1},\ldots,b_{n+p}$ such that $b_i=g_i(b_n,\ldots,b_{i-1})$.

Now consider the case $\underset{n}{\exists}\mathrm{UA(M)}\,(a_1,\ldots,a_{n-1})=+$. Then there exists an element a in E such that $\mathrm{UA(M)}\,(a_1,\ldots,a_{n-1},a)=+$, so that by assumption the condition $\mathscr{C}(\mathrm{U, A, M}, a_1,\ldots,a_{n-1},a)$ holds. Hence, for the index n there exists a function g_n over empty sequences, whose sole value is a. For every $\exists$-index i of U $(n+1\leqslant i\leqslant n+p)$, there exists a function f_i satisfying $\mathscr{C}_1$ and $\mathscr{C}_2$. With this function we associate the function g_i over the indices $n, n+1,\ldots,i-1$ assuming the same values as f_i, with n as an inactive index. Then condition $(\mathscr{C}_1)$ holds for the indices $n, n+1,\ldots,$

$i-1$ and $(\mathscr{C}_2)$ holds for b_n = value of $g_n = a$ and

$$b_i = g_i(b_n, b_{n+1}, \ldots, b_{i-1}) = f_i(b_{n+1}, \ldots, b_{i-1})$$

for $n+1 \leqslant i \leqslant n+p$.

5.6.2. *Denumerable-Model Theorem*

Assuming the axiom of choice:

For any multirelation M *with infinite base* E, *there exists a denumerable subset* D *of* E *such that the restriction* M | D *is logically equivalent to* M. More generally, *for any infinite subset* D *of* E, *there exists* D*(D ⊆ D* ⊆ E) *such that* M | D* *is logically equivalent to* M.

▷ We first consider a bound logical formula P such that P(M) = +. Expressing P in canonical prenex form UA, we get UA(M) = + and, by 5.6.1 above, the condition $\mathscr{C}$(U, A, M) holds with the empty sequence for $a_1, \ldots, a_n$. Now let D be a denumerable subset of E which is closed under each of the functions f_i. Then we have the condition $\mathscr{C}$(U, A, M | D), and therefore P(M | D) = +. If now P(M) = −, then ¬P(M) = +, therefore ¬P(M | D) = + and P(M | D) = −.

Now let P_j ($j = 1, 2, \ldots$) be the sequence of all bound formulas. For each one we have a canonical prenex formula $\mathrm{U}_j\mathrm{A}_j$, which is equideducible from either P_j (if P_j(M) = +) or ¬P_j (if P_j(M) = −) and such that the condition $\mathscr{C}$(U_j, A_j,M) holds for a finite number of functions f. Now let D be a denumerable subset of E which is closed under each of the functions f (the set of functions f is denumerable when j ranges over the natural numbers). Hence the conditions $\mathscr{C}$(U_j, A_j, M | D) hold, so that P_j(M | D) = P_j(M) for $j = 1, 2, \ldots$; this implies that M | D and M are logically equivalent. ◁

5.6.3. Any intersection of logical classes is a union of logical equivalence classes, i.e., is closed under logical equivalence. Now the class $\mathscr{A}$ of finite sets is closed under logical equivalence; thus, by 5.5.2, the intersection $\mathscr{A}'$ of the logical classes containing $\mathscr{A}$ is the class of all sets. We see in the same way that, if $\mathscr{A}$ is the class of finite sets of even cardinality, then $\mathscr{A}'$ consists of the sets of even cardinality and the infinite sets.

5.6.4. If $\mathscr{A}$ is a logical equivalence class and $\mathscr{P}_1, \ldots, \mathscr{P}_h$ a finite sequence of logical operators, then *the inverse image of* $\mathscr{A}$ *under* $\mathscr{P}_1, \ldots, \mathscr{P}_h$ *is an intersection of logical classes, hence a union of logical equivalence classes.* This is a consequence of 5.4.2 and of the fact that the equivalence class of M is the intersection of the logical classes containing M. This *inverse image* does not necessarily reduce to a logical equivalence class. For example, let $\mathscr{A}$ be the class of infinite unary relations which are always $+$, and $\mathscr{P}$ the operator which associates with each unary relation R the unary relation with the same base as R which is always $+$; then the inverse image is the class of all infinite unary relations.

Problem. Is every intersection of logical classes of infinite relations, the inverse image of a logical equivalence class under a logical operator?

Let $\mathscr{A}$ be a logical equivalence class consisting of μ-ary multirelations, and $\mathscr{P}$ a logical operator of predicarity μ; then *the image of $\mathscr{A}$ under $\mathscr{P}$ consists of logically equivalent relations, but it is not necessarily a logical equivalence class, even if $\mathscr{P}$ is free.* An example will be given in Volume 2, Chapter 1, Exercise 4, in connection with the class of logical equivalents of (I, A), where I is a chain, the sum of the chain of natural numbers and the chain of integers, and A a unary relation equal to $+$ for exactly one integer; the operator $\mathscr{P}$ is the selector which associates with any birelation its first relation. Then the resulting class contains I but does not contain the chain of natural numbers, which is logically equivalent to I.

5.7. Embedding logical class; compactness

The universal classes defined in 4.8 are special cases of logical classes: it suffices to take the free n-ary formula defining the class according to 4.8 and to prefix a quantifier $\forall_n^0$ reducing its arity n to 0 (if the base is empty, we must choose the quantifier $\forall_n^0$ that transforms the n-ary empty relation into $(\emptyset, +)$; see 3.6.2).

Recall that the union and intersection of two universal classes are universal classes, but the complement of a universal class need not be universal (4.8.1). We define an *embedding logical class* to be any finite union of finite intersections of universal classes and their complements. In other words, these are the classes obtained by closing the class of universal classes under the Boolean operations.

5.7.1. By 4.8.4, a class $\mathscr{A}$ is an embedding logical class if and only if one of the following equivalent conditions holds:

(1) *There exists an integer n such that if* R *is in* $\mathscr{A}$ *then any multirelation whose restrictions to subsets of* 0, 1,..., *n elements coincide with those of* R, *up to isomorphism, is also in* $\mathscr{A}$.

(2) $\mathscr{A}$ *is a finite union of classes* $\mathscr{B}$, *each defined by two finite sequences of finite multirelations* U *and* V, *such that* R *is in* $\mathscr{B}$ *if and only if every* U *is embeddable in* R *but no* V *is embeddable in* R.

As a particular case, we get the logical class of isomorphic images of a finite multirelation (see 5.4): in fact, R is isomorphic to a multirelation U with a finite base if U is embeddable in R but no extension of U to the set obtained by adding one element to the base is embeddable in R.

5.7.2. *Let* $\mathscr{A}$ *be an embedding logical class. If* $\mathscr{A}$ *contains a multirelation* R *with infinite base, there exists a finite subset* F *of the base of* R *such that any restriction of* R *to a superset of* F *is in* $\mathscr{A}$. *Moreover, the cardinality of* F *does not exceed a certain integer determined by the class.*

▷ Let n be an integer satisfying 5.7.1 (1), and, for every restriction of R to 1, 2,..., n elements, take an isomorphic restriction. This gives only a finite number of subsets of the base, with at most n elements: let F be the union of these subsets, which is finite. It follows from 5.7.1 that any restriction of R to a superset of F is in $\mathscr{A}$. Moreover, the number of possible nonisomorphic restrictions is determined by the arity and the integer n. ◁

Corollary: Given an infinite sequence of multirelations R_i $(i=0, 1, 2, \ldots,)$, belonging to $\mathscr{A}$ and such that R_{i+1} is an extension of R_i for any i, *the common extension of* R_i*'s to the union of bases, belongs to* $\mathscr{A}$.

5.7.3. In the particular case of embedding logical classes, the following compactness theorem is valid:

COMPACTNESS. *Given a set of embedding logical classes* $\mathscr{A}$ *with empty intersection, there exists a finite subset of classes* $\mathscr{A}$ *having empty intersection.*

▷ Since the set of classes $\mathscr{A}$ is denumerable, we can index them $\mathscr{A}_i$. Suppose that no finite intersection is empty. We can then replace each class by its intersection with its predecessors, that is, we may assume that the $\mathscr{A}_i$ form a decreasing sequence. For every integer i, there exist multirelations in $\mathscr{A}_i$ with arbitrarily large finite bases; for otherwise the multirelations in $\mathscr{A}_i$ would constitute a finite set (up to isomorphism), and the assertion of the theorem would follow immediately. We can stipulate that a certain finite set F_1 of integers, whose cardinality depends only on $\mathscr{A}_1$, be contained in all the bases of these multirelations, and moreover that the restrictions to F_1 and all its supersets belong to $\mathscr{A}_1$. Finally, since the number of possible restrictions is finite, we can also stipulate that all the above-mentioned multirelations have the same restriction to F_1, say R_1. Thus, for every integer i there exist multirelations in $\mathscr{A}_i$ with arbitrarily large bases, all extensions of R_1, all of whose restrictions to supersets of F_1 belong to $\mathscr{A}_1$. Repeating the argument, we get an extension R_2 of R_1, etc., and finally a common extension R of $R_1, R_2, \ldots$, defined on the set of integers. For each integer n, by the corollary of 5.7.2, R belongs to any $\mathscr{A}_n$. $\triangleleft$

5.7.4. Given a multirelation R, the intersection of all embedding logical classes containing R is the age of R, i.e., the class of multirelations having the same finite restrictions as R (up to isomorphism) (see 3.2.4).

An age may or may not be a logical equivalence class. For example, the age of infinite sets is a logical equivalence class (see 5.6). On the other hand, the set of chains with infinite bases is an age which is not a logical equivalence class, for it contains (among other things) chains with minimal elements and chains without minimal elements, and while the former satisfy the formula $\underset{x}{\exists}\underset{y}{\forall}\rho xy$, the latter do not.

EXERCISES

1

(1) Consider the following two formulas, where α is a unary predicate:

(i) $\alpha x \Leftrightarrow \underset{y}{\forall} \neg \alpha y$;

(ii) $\underset{y}{\forall} (\alpha y \Rightarrow y \equiv x)$.

Show that the negation of (ii) is deducible from (i). Note that if αx states that "x is true", then (i) states that "x is true if and only if everything is false", and (ii) states that "the only thing that can be true is x". Thus the conjunction of (i) and (ii) states: "the only thing that can be true is that everything is false".

(2) Retain formula (i), and replace (ii) by the following formula (ii'), where α and β are two unary predicates (β is inactive in (i) and (ii)):

(ii') $\underset{y}{\forall} (\alpha y \wedge \beta y) \Rightarrow y \equiv x$.

Show that (ii') is deducible from (ii), but not conversely. (Replace x by 1 and α, β by two relations A, B defined on $\{1, 2\}$ that assign (ii) the value $-$ and (ii') the value $+$.)

Show that the conjunction of (i) and (ii') is consistent. More precisely, it assumes the value $+$ when x is replaced by 1 and α, β by two suitable relations over $\{1, 2\}$.

Show that

$$\underset{y}{\exists} \alpha y \wedge \neg \beta y \wedge y \not\equiv x$$

is deducible from the conjunction of (i) and (ii'), but neither βx nor its negation is so deducible.

Note that if αx states that "x is true" and βx that "I say x", then the conjunction of (i) and (ii) states: "the only thing that I can say and that can be true is that everything is false". Hence we deduce: "there exists a truth which I do not say, and it is not: everything is false". We cannot deduce whether "I say that everything is false" or whether "I do not say that everything is false".

2

Consider the three following formulas, where α and β are unary predicates:

(i) $\alpha x \Leftrightarrow \underset{y}{\forall} (\beta y \Rightarrow \neg \alpha y)$;

(ii) $\underset{y}{\forall} (\alpha y \wedge \beta y) \Rightarrow y \equiv x$;

(iii) βx.

(1) Show that $\neg \alpha x$ is deducible from the conjunction of (i) and (iii); hence conclude that the conjunction of (i), (ii), (iii) is an antithesis.

(2) Show that the conjunctions of (i) and (ii), (ii) and (iii), (i) and (iii) are consistent (use relations over one or two elements).

(3) Show that αx is deducible from (i) and (ii).

If αx states that "x is true" and βx that "I say x", then $\underset{y}{\forall} (\beta y \Rightarrow \neg \alpha y)$ states that "I always tell lies" and the conjunction of (i), (ii), (iii) states that "I say that I always tell lies; this is the only truth that I can say."

3

Consider the following formulas, where α, β, γ are unary predicates:

(i) $\alpha x \wedge (\gamma x \Leftrightarrow \forall_z (\beta z \Rightarrow \neg \gamma z))$;

(ii) $\beta y \wedge (\gamma y \Leftrightarrow \exists_z (\alpha z \wedge \gamma z))$.

(1) Show that $\neg \gamma x$ is deducible from the conjunction of (i) and (ii); hence, so are $\exists_z \beta z \wedge \gamma z$ and $\exists_z \alpha z \wedge \neg \gamma z$.

(2) Show that the conjunction of (i) and (ii) is consistent, and that neither $\forall_z \alpha z \Leftrightarrow \beta z$ nor its negation are deducible from it (consider models over two or three elements).

Note that if we replace α by β and γ by α, then (i) is simply the conjunction of formulas (i) and (iii) of the preceding exercise.

If αx states that "the officer says x", βx that "the prisoner says x", and γx that "x is true", then (i) states: "the officer says: everything the prisoner says is false", and (ii) states: "the prisoner says: there is something true in what the officer says." Hence we deduce: "there is something true in what the prisoner says, and something false in what the officer says" (see [J. COH, 1957]).

4

(1) Show that the following two formulas are theses:

(i) $\exists_y (\rho xy \wedge \rho yy) \vee \forall_z \exists_t \neg \rho tz$,

(ii) $\forall_z (\rho zz \vee \exists_t \neg \rho tz)$.

For the first, consider the relation R substituted for the predicate, and distinguish between two cases: (a) there exists y such that $R(x, y) = +$ for all x; (b) there is no such y.

(2) Taking the conjunction of (i) and (ii), show that the first of the following formulas is a thesis; then use this to show that the other two formulas are theses:

$$\exists_y \forall_z \exists_t (\rho xy \wedge \rho yy \wedge \rho zz) \vee \neg \rho tz,$$
$$\exists_y \forall_z \exists_t (\rho xy \wedge \rho yt \wedge \rho zz) \vee \neg \rho tz,$$
$$\exists_x \forall_y \exists_z (\rho xx \wedge \rho yy \wedge \rho xz) \vee \neg \rho zy.$$

(3) Show that none of the formulas obtained from the above by changing any $\exists$ into a $\forall$ is a thesis.

5

The concept of finite set was defined in Chapter 1, Exercise 1. A (μ, n)-ary logical formula will be called a *finitary thesis* if it assumes the value $+$ for any μ-ary multirelation with finite base and any n-tuple of elements of this base.

(1) Let P be a bound formula with a binary predicate ρ, whose models are chains (total orderings). Let Q be the formula $\exists_x \forall_y \rho yx$ (there exists a maximal element). Show that $P \Rightarrow Q$ is a finitary thesis.

(2) Let P′ be the conjunction of $\forall_x \neg \rho xx$ (irreflexivity), $\forall_x \exists_y \rho xy$ (each element has a successor) and $\forall_{xx'y} (\rho xy \wedge \rho x'y) \Rightarrow x \equiv x'$ (each element has at most one predecessor). Let Q′ be the formula $\forall_y \exists_x \rho xy$ (each element has a predecessor). Show that $P' \Rightarrow Q'$ is a finitary thesis.

(3) Show that neither of the above two finitary theses is a thesis (provided that ordinary arithmetic is assumed to be consistent, so that the chain of integers has no maximal element

and the successor relation over the integers satisfies P′ but not Q′). Under the same assumptions, show that neither of the two finitary theses $P \Rightarrow Q$ and $P' \Rightarrow Q'$ is deducible from the other.

(4) Show that the following formulas are finitary theses but not theses:

$$\underset{x}{\exists}\underset{y}{\forall}\underset{z}{\exists}(\rho yz \wedge \neg \rho xz) \vee \neg \rho xx \vee \rho yx$$
$$\underset{x}{\exists}\underset{y}{\forall}\underset{z}{\exists}(\rho yz \wedge \neg \rho xz) \vee \neg \rho xy \vee \rho yy.$$

(5) Show that for formulas with only unary predicates the concepts of finitary thesis and thesis are equivalent (use 5.5.3 (2)).

6

A free formula with indices $x, y, \ldots$ is said to be *permuting* if, when prefixed by finite sequences of quantifiers $\forall$ and $\exists$ differing only in their order, the results are always equideducible logical formulas. For example, any free thesis or antithesis is permuting. The formula $\rho x \vee \rho y$ is permuting, for the formulas obtained by prefixing $\underset{x}{\forall}\underset{y}{\exists}$ and $\underset{y}{\exists}\underset{x}{\forall}$ are equideducible. The formula $\neg \rho x \vee \rho y \vee x \equiv y$ is permuting: prefixing $\underset{y}{\exists}\underset{x}{\forall}$ or $\underset{x}{\forall}\underset{y}{\exists}$, we get a thesis. On the other hand, neither $\rho x \Leftrightarrow \rho y$ nor $\rho x \vee \rho y \vee x \equiv y$ is permuting.

(1) Recalling Chapter 2, Exercise 1, let us consider permuting connective formulas. Let us say that a free formula has *separable indices* if it is obtained from a permuting connective formula with atoms $a, b, \ldots$ by replacing a by a free formula in which exactly one index x is active, b by a free formula in which exactly one (other) index y is active, etc. Show that a free formula with separable indices is permuting.

Examples: $\rho x \vee \sigma x \vee \rho y$, obtained from $a \vee b$ when a is replaced by $\rho x \vee \sigma x$ and b by ρy. Any thesis is obtained from an atom a, when a is replaced by the thesis itself (in which all the indices are inactive).

(2) Any free formula which is equideducible from a formula with separable indices is permuting. Example: $\neg \rho x \vee \rho y \vee x \equiv y$ does not have separable indices, but it is equideducible from $\neg \rho x \vee \rho y$, which has separable indices.

Problem. Is any permuting free formula with unary predicates equideducible from a free formula with separable indices? Note that this is false for formulas with binary predicates: for example, $\neg \rho x \vee \rho y \vee \sigma xy$ is not equideducible from any formula with separable indices, but it is permuting, since when prefixed with $\underset{y}{\exists}\underset{x}{\forall}$ or $\underset{x}{\forall}\underset{y}{\exists}$ it yields a thesis.

CHAPTER 6

COMPLETENESS AND INTERPOLATION THEOREMS

6.1. INSTANCE, DEPLOYMENT, DEPLOYED FORM, RETURN NUMBER, RENEWAL RANK

Consider a canonical prenex formula P=UA, where A is a free formula (the *free part* of P) and U the *prefix* of P – a sequence of quantifiers whose indices assume consecutive increasing values (see 5.1). Let n be the arity of P, and $n+p$ that of A, so that $1, 2, \dots, n$ are the *free indices*, and $n+1, \dots, n+p$ the *bound indices* of P.

6.1.1. We define an *instance* of P to be any free formula $\mathrm{A}]f$ obtained from the free part A by an increasing mapping f whose restriction to $1, \dots, n$ is the identity (see 4.6.5); in other words,

$$f(1)=1, \quad \dots, \quad f(n)=n<f(n+1)<\cdots<f(n+p).$$

For every integer x $(1\leqslant x\leqslant n+p)$, the number $y=f(x)$ will be called a *value* of the instance; the integer x is called the *rank* of y in the instance $\mathrm{A}]f$. Moreover, depending on whether x is an $\forall$-index or an $\exists$-index in the prefix U, the value $y=f(x)$ will be called an $\forall$-*value* or an $\exists$-*value* in the instance $\mathrm{A}]f$.

6.1.2. Consider a finite or denumerable sequence of functions $f_1, \dots, f_i, \dots$ all defined and increasing on $\{1, \dots, n, \dots, n+p\}$, whose restriction to $1, \dots, n$ is the identity. Thus, with each f_i we associate an instance $\mathrm{A}_i=\mathrm{A}]f_i$. Then the sequence of instances is called a *deployment* of the formula P, if the functions f_i satisfy the above conditions and moreover there exist two functions u and r with the following properties.

$r(i)$ is an integer, called the *renewal rank* of i, such that

$$r(1)=n+1 \quad \text{and} \quad n+1\leqslant r(i)\leqslant n+p$$

for all i. $u(i)$ is an integer, called the *return number* of i, defined for $i\geqslant 2$, such that $u(i) < i$. We set

$$f_i(x) = f_{u(i)}(x) \quad \text{for} \quad x = 1, \ldots, n, n+1, \ldots, r(i) - 1;$$

for the values $x = r(i)$, $r(i)+1, \ldots, n+p$, we stipulate that $f_i(x)$ be strictly greater than all values already taken by $f_1, f_2, \ldots, f_{i-1}$.

We see that the function r is uniquely determined by the f's: $r(i)$ is the first x from which $f_i(x)$ is different from the values of $f_j (j < i)$. This is not true for the function u; one possibility is to define $u(i)$ by returning to the *first* function $f_j (j < i)$ such that $f_j(x) = f_i(x)$ for $x < r(i)$.

With this definition, $r(u(i)) \leqslant r(i)$ *for all* $i \geqslant 2$. In fact, either $u(i) = 1$, and then

$$r(u(i)) = n + 1 \leqslant r(i),$$

or $u(i) \geqslant 2$, in which case $u(u(i))$ exists and

$$f_i(x) = f_{u(i)}(x) = f_{u(u(i))}(x)$$

for all $x < r(i)$ and $< r(u(i))$; were $r(u(i)) > r(i)$, the preceding equality would be true for $x < r(i)$, so that $u(i)$ could be replaced by the (strictly) smaller number $u(u(i))$.

Given a positive integer y, there may exist one or more i, but at most one x which is the rank of y in f_i (so that $f_i(x) = y$). We shall say that x is the *rank of* y *in the deployment.* If x is an $\forall$-index ($\exists$-index) in the prefix U, we shall call y an $\forall$*-value* ($\exists$*-value*) *in the deployment.*

6.1.3. Given a finite deployment $A_1, \ldots, A_h$, consider the disjunction $A_1 \bigvee \ldots \bigvee A_h$. Moreover, with each positive integer i ($1 \leqslant i \leqslant h$) let us associate the *partial prefix* U_i whose indices are the values $f_i(r(i)), \ldots, f_i(n+p)$, and such that the symbol $\forall$ or $\exists$ associated with each is the same as that associated with rank $r(i), \ldots, n+p$, respectively, in U. Then the prenex formula $U_1 \ldots U_h(A_1 \bigvee \ldots \bigvee A_h)$ is called a *deployed form* of P, and the number h is its *order.*

Example. We start with the canonical prenex formula

$$\underset{2\,3}{\exists\forall} \rho x^1 x^3 \bigvee x^2 \not\equiv x^3$$

with free index 1 and bound indices 2 and 3. Set

$$f_1(2) = 3, \qquad f_1(3) = 5$$

(so that, evidently, $r(1)=n+1=2$). Then

$$A_1 = A]f_1 = \rho x^1 x^5 \bigvee x^3 \not\equiv x^5.$$

Now the return number is $u(2)=1$ and the renewal rank $r(2)=2$, with $f_2(2)=6, f_2(3)=8$; therefore

$$A_2 = A]f_2 = \rho x^1 x^8 \bigvee x^6 \not\equiv x^8.$$

Now set $u(3)=1$ and $r(3)=3$, with $f_3(2)=f_1(2)=3, f_3(3)=9$; therefore

$$A_3 = A]f_3 = \rho x^1 x^9 \bigvee x^3 \not\equiv x^9.$$

We obtain the deployed form

$$\underset{3}{\exists}\underset{5}{\forall}\underset{6}{\exists}\underset{8}{\forall}\underset{9}{\forall} \rho x^1 x^5 \bigvee x^3 \not\equiv x^5 \bigvee \rho x^1 x^8 \bigvee x^5 \not\equiv x^8 \bigvee \rho x^1 x^9 \bigvee x^3 \not\equiv x^9.$$

6.1.4. *For any deployed form* P *of* UA, *there exists a canonical deployed form* P′ *of* UA, *of the same order as* P, *and* P *is the transform of* P′ *under an injective function whose restriction to* $1,\ldots,n$ *is the identity* (so that P is equideducible from P′).

▷ Let $A_1 = A]f_1, \ldots, A_h = A]f_h$ be the deployment defining P. Replace f_1 by f'_1 = identity for $1,\ldots,n+p$, so that $A'_1 = A]f'_1 = A$. Then, for every $i \geqslant 2$, we define the new values

$$f'_i(r(i)), \quad f'_i(r(i)+1), \quad \ldots, \quad f'_i(n+p)$$

as the smallest integer m not already a value of one of $f'_1, \ldots, f'_{i-1}$, and the consecutive integers following m. ◁

6.1.5. *Given a canonical prenex formula* P *and an integer h, there are only finitely many canonical deployed forms of* P *of order h.* Thus there are only finitely many deployed forms of order h, up to bijective transformations.

Thus we have an algorithm for deciding, given two prenex formulas P and P′, whether P′ is a deployed form of P or not. Convert both P and P′ into canonical formulas by bijective transformations. Let n be the arity of P, which we assume equal to that of P′, and let $n+p$ and $n+p'$ be the arities of the free parts of P and P′. The order of the (alleged) deployed form P′ is at most p'. By a finite number of trials, one can compare P′ with all the canonical deployed forms of P of order $\leqslant p'$.

6.2. Every deployed form of a formula P is equideducible from P

▷ The assertion is true for a deployed form of order 1, since such a deployed form is the transform of P under a bijective function which is the identity on the free indices (see 5.3.4).

Consider a deployed form of order $h \geqslant 2$, defined by instances $A_1, \ldots, A_{h-1}, A_h$ and prefixes $U_1, \ldots, U_{h-1}, U_h$. It will suffice to show that it is equideducible from the deployed form of order $h-1$ defined by $A_1, \ldots, A_{h-1}$ and $U_1, \ldots, U_{h-1}$.

Let $k = u(h)$ be the return number of h. We may assume without loss of generality that the renewal ranks satisfy the inequality $r(k) \leqslant r(h)$ (see 6.1.2). Write the prefix U_k as $U_k = VW$, where V corresponds to $r(k)$, $r(k)+1, \ldots, r(h)-1$, and W to $r(h), \ldots, n+p$, i.e., to the same ranks as U_h. Then WA_k and U_hA_h are transforms of each other under a bijective correspondence which reduces to the identity on $1, 2, \ldots, n$ (and even up to $f_h(r(h)-1)$). Hence we have the equideduction $WA_k \dashv\vdash U_hA_h$.

Write the deployed form of order h as $U_1 \ldots U_{k-1}VB$, where

$$B = WU_{k+1} \ldots U_{h-1}U_h(A_1 \vee \cdots \vee A_{k-1} \vee A_k \vee A_{k+1} \vee \cdots \vee A_{h-1} \vee A_h).$$

The indices of U_h are inactive in $A_1, \ldots, A_{h-1}$, and those of $U_{k+1}, \ldots, U_{h-1}$ are inactive in $A_1, \ldots, A_k$; thus

$$B \dashv\vdash W(A_1 \vee \cdots \vee A_k \vee U_{k+1} \ldots U_{h-1}(A_{k+1} \vee \cdots \vee A_{h-1} \vee U_hA_h)),$$

Set

$$C = U_{k+1} \cdots U_{h-1}(A_{k+1} \vee \cdots \vee A_{h-1} \vee U_hA_h).$$

Replacing U_hA_h by WA_k, and noting that the indices of $U_{k+1}, \ldots, U_{h-1}$ are inactive in A_k, we obtain

$$C \dashv\vdash WA_k \vee U_{k+1} \ldots U_{h-1}(A_{k+1} \vee \cdots \vee A_{h-1}).$$

It follows that $B \dashv\vdash W(A_1 \vee \cdots \vee A_k \vee C)$, so that

$$B \dashv\vdash W(A_1 \vee \cdots \vee A_{k-1} \vee A_k \vee WA_k \vee U_{k+1} \ldots U_{h-1}(A_{k+1} \vee \cdots \vee A_{h-1})).$$

We now apply 3.7.5, with $\mathscr{P}$ replaced by W, the relation R by A_k (or, more precisely, the transform under A_k of a given multirelation), and finally S by

$$A_1\bigvee\cdots\bigvee A_{k-1}\bigvee U_{k+1}\ldots U_{h-1}(A_{k+1}\bigvee\cdots\bigvee A_{h-1}),$$

thus getting the equideduction

$$B -\!\vdash W(A_1\bigvee\cdots\bigvee A_{k-1}\bigvee A_k\bigvee U_{k+1}\ldots U_{h-1}(A_{k+1}\ldots A_{h-1})),$$

and finally

$$B -\!\vdash WU_{k+1}\ldots U_{h-1}(A_1\bigvee\cdots\bigvee A_{h-1}).$$

It follows that the deployed form $U_1\ldots U_{k-1}VB$ is equideducible from the deployed form of order $h-1$:

$$U_1\ldots U_{h-1}(A_1\bigvee\cdots\bigvee A_{h-1}).\ \lhd$$

6.2.1. In the example given in 6.1, the formula $\underset{2\,3}{\exists\forall}\rho x^1x^3\bigvee x^2\not\equiv x^3$ is equideducible from the following formulas:

$\underset{3\,5}{\exists\forall}\rho x^1x^5\bigvee x^3\not\equiv x^5$,

$\underset{3\,5}{\exists\forall}(\rho x^1x^5\bigvee x^3\not\equiv x^5\bigvee\underset{3\,5}{\exists\forall}(\rho x^1x^5\bigvee x^3\not\equiv x^5))$,

$\underset{3\,5}{\exists\forall}(\rho x^1x^5\bigvee x^3\not\equiv x^5\bigvee\underset{6\,8}{\exists\forall}(\rho x^1x^8\bigvee x^6\not\equiv x^8))$,

$\underset{3\,5}{\exists\forall}(\rho x^1x^5\bigvee x^3\not\equiv x^5\bigvee\underset{5}{\forall}(\rho x^1x^5\bigvee x^3\not\equiv x^5)\bigvee\underset{6\,8}{\exists\forall}(\rho x^1x^8\bigvee x^6\not\equiv x^8))$,

$\underset{3\,5}{\exists\forall}(\rho x^1x^5\bigvee x^3\not\equiv x^5\bigvee\underset{9}{\forall}(\rho x^1x^9\bigvee x^3\not\equiv x^9)\bigvee\underset{6\,8}{\exists\forall}(\rho x^1x^8\bigvee x^6\not\equiv x^8))$,

$\underset{3\,5\,6\,8\,9}{\exists\forall\exists\forall\forall}\rho x^1x^5\bigvee x^3\not\equiv x^5\bigvee\rho x^1x^8\bigvee x^6\not\equiv x^8\bigvee\rho x^1x^9\bigvee x^3\not\equiv x^9$.

6.3. COMPLETENESS THEOREM ([HER, 1930], and in a less elaborate form, [GOD, 1930]; see also [HEN, 1949a])

A canonical prenex formula is a thesis if and only if one of its deployed forms is an explicit thesis.

The condition is sufficient – this follows from the fact that any explicit thesis is a thesis (5.3.1) and any deployed form of a formula is equideducible from it (6.2).

We shall prove necessity in Sections 6.4 and 6.5 below.

In other words, by taking each explicit thesis together with the formulas of which it is a deployed form, we obtain all possible theses.

6.3.1. *Examples of Deployed Theses*

(1) The formula $\underset{1\,2}{\exists\forall}\rho x^1 \vee \neg\rho x^2$ is a thesis. The following deployed form is an explicit thesis:

$$\underset{1\,2\,3\,4}{\exists\forall\exists\forall}\rho x^1 \vee \neg\rho x^2 \vee \rho x^3 \vee \neg\rho x^4 .$$

Taking $h(3)=2$, we get the free thesis

$$\rho x^1 \vee \neg\rho x^2 \vee \rho x^2 \vee \neg\rho x^4 .$$

(2) The formula

$$\underset{1\,2\,3}{\forall\exists\forall}\rho x^1 x^2 \vee \neg\rho x^1 x^3$$

is a thesis; we get the explicit thesis

$$\underset{1\,2\,3\,4\,5}{\forall\exists\forall\exists\forall}\rho x^1 x^2 \vee \neg\rho x^1 x^3 \vee \rho x^1 x^4 \vee \neg\rho x^1 x^5 ,$$

where $h(4)=3$.

(3) The formula

$$\underset{1\,2\,3}{\exists\forall\forall}\neg\rho x^1 \vee (\rho x^2 \wedge \rho x^3)$$

is a thesis; we get the explicit thesis

$$\underset{1\,2\,3\,4\,5\,6\,7\,8\,9}{\exists\forall\forall\exists\forall\forall\exists\forall\forall}$$

$$\neg\rho x^1 \vee (\rho x^2 \wedge \rho x^3) \vee \neg\rho x^4 \vee (\rho x^5 \wedge \rho x^6) \vee \neg\rho x^7 \vee (\rho x^8 \wedge \rho x^9),$$

where

$$h(4)=2, \qquad h(7)=3 .$$

(4) The formula

$$\underset{1\,2\,3\,4}{\exists\forall\exists\exists}\neg\rho x^1 x^2 \vee (\rho x^2 x^3 \wedge \rho x^3 x^4)$$

is a thesis; the following deployed form is an explicit thesis (with the instances and corresponding partial prefixes written below each other):

$$\underset{1}{\exists}\underset{2}{\forall}\underset{3}{\exists}\underset{4}{\exists}\quad \neg\rho x^1x^2 \bigvee(\rho x^2\, x^3 \bigwedge\rho x^3\, x^4\,),$$

$$\underset{5}{\exists}\underset{6}{\forall}\underset{7}{\exists}\underset{8}{\exists}\bigvee\neg\rho x^5x^6 \bigvee(\rho x^6\, x^7 \bigwedge\rho x^7\, x^8\,),$$

$$\underset{9}{\exists}\underset{10}{\forall}\underset{11}{\exists}\underset{12}{\exists}\bigvee\neg\rho x^9x^{10}\bigvee(\rho x^{10}x^{11}\bigwedge\rho x^{11}x^{12}),$$

$$\underset{13}{\exists}\underset{14}{\exists}\bigvee\neg\rho x^1x^2 \bigvee(\rho x^2\, x^{13}\bigwedge\rho x^{13}x^{14}).$$

Setting $h(5)=2$, $h(9)=h(13)=6$, $h(14)=10$, we get the free thesis:

$$\neg\rho x^1x^2\bigvee\cdots\bigvee\neg\rho x^2x^6\bigvee\cdots\bigvee\neg\rho x^6x^{10}\bigvee\cdots\bigvee\,(\rho x^2x^6\bigwedge\rho x^6x^{10}).$$

6.3.2. The completeness theorem furnishes an enumeration algorithm for theses. We first enumerate the explicit theses, which is possible since we have an algorithm deciding whether a given logical formula is an explicit thesis or not (5.3.2). Then, for each explicit thesis, P say, we enumerate the (finite number of) formulas of which P is a deployed form (see 6.1.5). We have already mentioned (see 5.3.2) that in all probability there exists no algorithm deciding whether a logical formula is a thesis or not.

6.4. Lemmas on Deployment

In the deployment of $\mathrm{P}=\mathrm{UA}$, we shall say that the instance A_k *returns transitively* to $\mathrm{A}_i(i\leqslant k)$ if i is a term in the finite decreasing sequence $k, u(k), u(u(k)),\ldots$. This relation between i and k depends on the specific function u; nevertheless:

6.4.1. If u is the minimum function compatible with the functions f (see 6.1.2), and A_i is the first instance which takes a given value u_0 at rank $r(i)$, then *the instances which return transitively to* A_i *are exactly those which take the value* u_0 *at rank* $r(i)$.

6.4.2. *Let* $\mathrm{P}=\underset{n+1}{\forall}\mathrm{Q}$ *or* $\underset{n+1}{\exists}\mathrm{Q}$, *where*

$$\mathrm{Q}=\underset{n+2}{\mathrm{U}}\cdots\underset{n+p}{\mathrm{U}}\mathrm{A}(x^1,\ldots,x^n,x^{n+1},\ldots,x^{n+p}),$$

and let $\mathscr{A}$ *be a deployment of* P. *Then the instances of* $\mathscr{A}$ *in which* $n+1$ *is mapped onto a given integer* $u\geqslant n+1$ *constitute a deployment of*

$$\mathrm{Q}_u = \mathrm{Q}]^{u}_{n+1}{}^{u+1}_{n+2}\cdots{}^{u+p-1}_{n+p}$$
$$= \underset{u+1}{\mathrm{U}_2}\cdots\underset{u+p-1}{\mathrm{U}_p}\,\mathrm{A}\,(x^1,\ldots,x^n,x^u,x^{u+1},\ldots,x^{u+p-1}).$$

▷ Consider the first instance of $\mathscr{A}$ in which $n+1$ is mapped onto u, say $\mathrm{A}_1 = \mathrm{A}(x^1,\ldots,x^n,x^u,x^{n_2},\ldots,x^{n_p})$, with $u<n_2<\cdots<n_p$. This is the instance of Q_u defined by the function f which is the identity on $1,\ldots,u$ and

$$f(u+1)=n_2,\ldots,f(u+p-1)=n_p.$$

The instances in which $n+1$ is mapped onto u are exactly those which return transitively to A_1 (see 6.4.1). They satisfy the conditions of 6.1.2, with the renewal ranks $n+1,\ldots,n+p$ replaced by $u,\ldots,u+p-1$. ◁

6.4.3. *Let $\mathscr{A}$ be a deployment of* P *and consider two integers, $u \leqslant$ arity of* P *and $v<u$. Then the instances of* A *as transformed by $|^v_u$ constitute a deployment of* $\mathrm{P}]^v_u$ (notation as defined in 4.6.5 and 5.3.4).

6.5. Completed deployment, bi-instance

Given a canonical prenex formula P, we define a *completed deployment* as a pair $(\mathscr{A}, h)$, where $\mathscr{A}$ is a deployment of P and h a function, the *completing function*, defined as follows on the values of $\mathscr{A}$, with positive integer values. The function h is the identity for $1,\ldots,n$ and for all ∀-values; $h(i)\leqslant i$ for each ∃-value. Moreover, for every instance $\mathrm{A}\in\mathscr{A}$, all q $(1\leqslant q\leqslant p)$ such that $n+q$ is an ∃-index, and all integers v, there exists an instance $\mathrm{B}\in\mathscr{A}$, defined by a function f_{B}, say, taking the same values as A at ranks $1,2,\ldots,n+q-1$ and at rank $n+q$ a value $u=f_{\mathrm{B}}(n+q)$ mapped by h onto v; in other words,

$$h(u)=hf_{\mathrm{B}}(n+q)=v.$$

For every instance A, the transform $\mathrm{A}]h$ will be called a *bi-instance* of the completed deployment.

6.5.1. *Every canonical prenex formula has a completed deployment.*

▷ Define f_1 as the identity on $1,\ldots,n,\ldots,n+p$. For every $i\geqslant 2$ and all q $(1\leqslant q\leqslant p)$ such that $n+q$ is associated with ∃, take an infinite set of integers $j>i$ such that $u(j)=i$ and $r(j)=n+q$. Finally, for each sequence of values of a function f_i at ranks $1,\ldots,n,\ldots,n+q-1$, we consider the (infinitely many) values y at rank $n+q$, and let $h(y)$ range over all the positive integers. ◁

6.5.2.1. *Let* $P = \underset{n+1}{\forall} Q$ or $\underset{n+1}{\exists} Q$, *where*

$$Q = \underset{n+2}{U_2}\cdots\underset{n+p}{U_p} A(x^1,\ldots,x^n,x^{n+1},x^{n+2},\ldots,x^{n+p}),$$

let $(\mathscr{A}, h)$ *be a completed deployment of* P *and* u *an integer such that* $\mathscr{A}$ *contains a deployed instance which takes the value* u *at rank* $n+1$. *Then there exists a completed deployment* $(\mathscr{B}, l)$ *of the following formula:*

$$Q_{u,h} = \underset{u+1}{U_2}\cdots\underset{u+p-1}{U_p} A(x^1,\ldots,x^n,x^{h(u)},x^{u+1},\ldots,x^{u+p-1})$$

in which each bi-instance is a bi-instance of $(\mathscr{A}, h)$.

▷ Consider the first instance in $\mathscr{A}$ which takes the value u at rank $n+1$; let this be $A(x^1,\ldots,x^n,x^u,x^{n_2},\ldots,x^{n_p})$, where

$$n+1 \leqslant u < n_2 < \cdots < n_p.$$

Let $\mathscr{B}'$ be the sequence extracted from $\mathscr{A}$, consisting of all the instances which take the value u at rank $n+1$: these are the instances which return transitively to the first (see 6.4.1). Let $\mathscr{B}$ be the sequence obtained from $\mathscr{B}'$ by subjecting each of its instances to the substitution $|_u^{h(u)}$, i.e., replacing the value u at rank $n+1$ by $h(u)$. Moreover $\mathscr{B}'$ is a deployment of

$$\underset{u+1}{U_2}\cdots\underset{u+p-1}{U_p} A(x^1,\ldots,x^n,x^u,x^{u+1},\ldots,x^{u+p-1}) \quad \text{(see 6.4.2)}$$

and $\mathscr{B}$ is a deployment of the transform of this formula under $|_u^{h(u)}$, i.e., a deployment of the formula $Q_{u,h}$ as required (see 6.4.3).

Define the function l as the identity on $\{1,\ldots,u\}$ and $l(y)=h(y)$ for $y \geqslant u+1$, and consider the system $(\mathscr{B}, l)$.

(1) Any instance B in $\mathscr{B}$ originates from an instance B′ in $\mathscr{B}'$, therefore also in $\mathscr{A}$, by the transformation $B = B']_u^{h(u)}$. Thus the bi-instance $B]l = B']_u^{h(u)}]l$ is precisely $B']h$, with the ranks $1,\ldots,n,u,u+1,\ldots,u+p-1$ transformed into

$$1,\ldots,n,h(u),h(u+1)=l(u+1),\ldots,h(u+p-1)=l(u+p-1).$$

(2) l is a completing function, and consequently $(\mathscr{B}, l)$ is a completed

deployment of $Q_{u,h}$. Indeed, let the instance B be in $\mathscr{B}$, originating from B′ as before, and let q $(2 \leqslant q \leqslant p)$ be an integer such that $n+q$ is an ∃-index of P; thus $u+q-1$ is an ∃-index of $Q_{u,h}$. Given any positive integer z, by assumption $\mathscr{A}$ contains an instance C′ taking the same values as B′ at all ranks $\leqslant n+q-1$ (relative to P), and a value y at rank $n+q$ such that $h(y)=z$. In particular, the value of C′ at rank $n+1$ is u, and therefore it belongs to $\mathscr{B}'$. Its transform $C = C']_u^{h(u)}$ belongs to $\mathscr{B}$. Moreover, since the value increases with rank, $n+1<n+q$ implies $u<y$. Thus, the formulas B′ and C′, hence also B and C, take the same value at all ranks $\leqslant n+q-1$ (relative to P) or $\leqslant u+q-2$ (relative to $Q_{u,h}$) and C takes at rank $n+q$ (or $u+q-1$) the value y such that $l(y)=h(y)=z$. Thus the deployment $\mathscr{B}$ is completed by l, and the assertion is proved. ◁

6.5.2.2. *Let* $P = \underset{n+1}{\forall} Q$, *and let* $(\mathscr{A}, h)$ *be a completed deployment of* P. *Then there exist an integer* $u \geqslant n+1$ *and a completed deployment* $(\mathscr{B}, l)$ *of the formula*

$$Q'_u = \underset{u+1}{U_2} \ldots \underset{u+p-1}{U_p} A(x^1, \ldots, x^n, x^u, x^{u+1}, \ldots, x^{u+p-1})$$

in which every bi-instance is a bi-instance of $(\mathscr{A}, h)$.

This follows from 6.5.2.1, by letting u be the value at rank $n+1$ in the first instance of $\mathscr{A}$, with $h(u)=u$ since the quantifier of rank $n+1$ is ∀.

6.5.2.3. *Let* $P = \underset{n+1}{\exists} Q$, *and let* $(\mathscr{A}, h)$ *be a completed deployment of* P. *Then, for every integer* v, *there exist an integer* $u \geqslant \max(n+1, v)$ *and a completed deployment* $(\mathscr{B}, l)$ *of the formula*

$$Q''_{u,v} = \underset{u+1}{U_2} \ldots \underset{u+p-1}{U_p} A(x^1, \ldots, x^n, x^v, x^{u+1}, \ldots, x^{u+p-1})$$

in which every bi-instance is a bi-instance of $(\mathscr{A}, h)$.

This follows from 6.5.2.1 and from the fact that, since $(\mathscr{A}, h)$ is a completed deployment, for every v there exists u such that (1) $h(u)=v$, and (2) there exists an instance in $\mathscr{A}$ which assumes the value u at rank $n+1$.

6.5.3. *Let* P *be a canonical prenex formula,* s *a sequence, and* M *a multi-relation over the set of elements of* s. *If the system* (s, M) *assigns the value* – *to all bi-instances of a completed deployment of* P, *then it assigns the value* – *to* P.

▷ The assertion is obvious if P is free, since then the instances and bi-instances all coincide with P. We now consider the general case in which P begins with $\underset{n+1}{\forall}$ or $\underset{n+1}{\exists}$.

(1) Let P have the form $\underset{n+1}{\forall} A(x^1,\ldots,x^n,x^{n+1})$, where A is free. Then every instance has the form $A(x^1,\ldots,x^n,x^u)$, where $u \geqslant n+1$, and the bi-instances coincide with the instances. Thus, if (s, M) gives these bi-instances the value $-$, it gives the value $-$ to a free formula $A(x^1,\ldots,x^n,x^u)$, therefore to P, since $P \vdash A(x^1,\ldots,x^n,x^u)$.

In general, let P have the form $\underset{n+1}{\forall} Q$, where

$$Q = \underset{n+2}{U_2} \ldots \underset{n+p}{U_p} A(x^1,\ldots,x^n,x^{n+1},x^{n+2},\ldots,x^{n+p}).$$

If (s, M) gives the value $-$ to all bi-instances of a completed deployment of P, then, by 6.5.2.2, there exist an integer u and a completed deployment of

$$Q'_u = \underset{u+1}{U_2} \ldots \underset{u+p-1}{U_p} A(x^1,\ldots,x^n,x^u,x^{u+1},\ldots,x^{u+p-1}),$$

such that (s, M) gives the value $-$ to all bi-instances of this completed deployment. By the induction hypothesis, (s, M) gives the value $-$ to Q'_u, therefore also to P, since $P \vdash Q'_u$.

(2) Let P have the form $\underset{n+1}{\exists} A(x^1,\ldots,x^n,x^{n+1})$. Then any instance has the form $A(x^1,\ldots,x^n,x^u)$, where $u \geqslant n+1$. Any completed deployment, will contain these instances for infinitely many values u, and also a completing function h which associates all integer values $v = h(u)$ to these values of u. Thus, for every v, we have the bi-instance $A(x^1,\ldots,x^n,x^v)$. If (s, M) gives the value $-$ to these bi-instances then it gives P the value $-$.

Now let P have the form

$$\underset{n+1}{\exists} Q, \text{ where } Q = \underset{n+2}{U_2} \ldots \underset{n+p}{U_p} A(x^1,\ldots,x^n,x^{n+1},x^{n+2},\ldots,x^{n+p}).$$

If (s, M) gives the value $-$ to all bi-instances of a completed deployment of P, then, by 6.5.2.3, there exist for every integer v a value u and a completed deployment of

$$Q''_{u,v} = \underset{u+1}{U_2} \ldots \underset{u+p-1}{U_p} A(x^1,\ldots,x^n,x^v,x^{u+1},\ldots,x^{u+p-1}),$$

where $u \geqslant \max(n+1, v)$, such that (s, M) gives the value $-$ to all the bi-instances of this completed deployment. By the induction hypothesis (s, M) gives the value $-$ to $\mathrm{Q}''_{u,v}$ for all v. Thus, for any value (from the sequence s) replacing the term of rank $n+1$, the system (s, M) gives Q the value $-$; hence (s, M) gives P the value $-$. $\lhd$

6.5.4. We can now prove the completeness theorem. Let P be a formula for which no deployed form is an explicit thesis. Let $(\mathscr{A}, h)$ be a completed deployment of P (see 6.5.1), and let $\mathscr{A} = \mathrm{A}_1, \ldots, \mathrm{A}_i, \ldots$. For each i, the transform $(\mathrm{A}_1 \bigvee \cdots \bigvee \mathrm{A}_i)]\, h$ is not a thesis; hence there exists a system (s_i, M_i) which gives it the value $-$. By the compactness lemma for free operators (see 4.4), there exists a system (s, M) which gives the value $-$ to all bi-instances $\mathrm{A}_i]\, h$, therefore also to P. Hence P cannot be a thesis.

6.5.5. Note that the completeness theorem implies a weakened version of the denumerable-model theorem [LOW, 1915]: *If there exist models of a logical formula* P, *then one of them is finite or denumerable.*

$\rhd$ If there exist multirelations which give P the value $+$, then $\neg$P is not a thesis. There exist a completed deployment of $\neg$P and, by 6.5.4, a system (s, M), *where the sequence s is denumerable* (though terms may be repeated, so that the *set* of terms of s may be finite), which gives the value $-$ to all bi-instances. Thus this system gives $\neg$P the value $-$, and so gives P the value $+$. $\lhd$

This weakened version of the denumerable-model theorem has been proved without the axiom of choice or the ultrafilter axiom.

6.6. Flexible deployed form

We now slightly modify the definition of a deployment (6.1.2), stipulating that the values of f_i (therefore also of $\mathrm{A}_i = \mathrm{A}]\, f_i$), beginning from the renewal rank $r(i)$, are distinct from but not necessarily greater than previously assumed values of $f_1, \ldots, f_{i-1}$. The prefix still contains the indices in increasing order (therefore, in decreasing order, as regards the order of application of the quantifiers). The resulting deployed form is called a *flexible deployed form.*

An equivalent definition is the following, *which makes no appeal to the order of the instances.*

(1) Every instance B is obtained by transforming the free part A of the original formula by an increasing function f, defined for $1, 2, \ldots, n+p$,

such that

$$f(1)=1, \ldots, f(n)=n<f(n+1)<\cdots<f(n+p).$$

(2) If two functions f and g assume the same value, they can only do so at the same rank x, and they then assume the same values for all ranks $\leqslant x$.

(3) The prefix of the deployed form is defined by stipulating that the quantifier associated with each value y of a function f at rank x is the same quantifier $\forall$ or $\exists$ associated with x; these quantifiers are arranged in the order of values.

Example. The following formula is a flexible deployed form of the formula $\underset{2\,3}{\exists\forall}\rho x^1x^3\bigvee x^2\not\equiv x^3$:

$$\underset{3\,4\,5\,6\,7}{\exists\exists\forall\forall\forall}\rho x^1x^5\bigvee x^3\not\equiv x^5\bigvee\rho x^1x^7\bigvee x^4\not\equiv x^7\bigvee\rho x^1x^6\bigvee x^4\not\equiv x^6,$$

where the functions f, denoted here by f, g, h, are defined on $\{1, 2, 3\}$ as: $f(1)=g(1)=h(1)=1, f(2)=3, f(3)=5, g(2)=4, g(3)=7, h(2)=4, h(3)=6.$

The role of the partial prefixes, in the sense of 6.1.3, proves to be unimportant here, since they depend on the order in which the instances are arranged (according to the first definition of flexible deployed form). Thus, in the above example with the order f, g, h we get partial prefixes (3, 5), (4, 7), (6); but if we take the order f, h, g, we get (3, 5), (4, 6), (7). On the other hand, with each instance, the transform of A by a function f, say, we can associate a unique *extracted prefix*, defined as the increasing sequence of indices $f(n+1), \ldots, f(n+p)$, each associated with the quantifier $\forall$ or $\exists$ of the corresponding rank $n+1, \ldots, n+p$ in $\mathrm{P}=\mathrm{UA}$. Every extracted prefix contains the same number p of quantifiers as U, but with another sequence of indices. In the above example, the extracted prefixes associated with the functions f, g, h are (3, 5), (4, 7), (4, 6), respectively.

6.6.1. The flexible deployed form is a more general concept than the strict deployed form of 6.1.3, if we adopt the first definition. However, this seeming generality is merely formal: *propositions* 6.2 *and* 6.3 *remain valid for flexible deployed forms.*

As an example, consider the following thesis:

$$\underset{1\,2\,3}{\exists\forall\forall}\neg\rho x^1x^2\bigvee(\rho x^2x^3\bigwedge x^2\not\equiv x^3)\bigvee\rho x^2x^2,$$

which has the following strict deployed form (among others):

$$\underset{1\,2\,3}{\exists\forall\exists}\quad \neg\rho x^1x^2\bigvee(\rho x^2x^3\bigwedge x^2\not\equiv x^3)\bigvee\rho x^2x^2,$$
$$\underset{4\,5\,6}{\exists\forall\exists}\bigvee\neg\rho x^4x^5\bigvee(\rho x^5x^6\bigwedge x^5\not\equiv x^6)\bigvee\rho x^5x^5,$$
$$\underset{7}{\exists}\bigvee\neg\rho x^1x^2\bigvee(\rho x^2x^7\bigwedge x^2\not\equiv x^7)\bigvee\rho x^2x^2.$$

This formula yields a free thesis when 4 is replaced by 2 and 7 by 5. But it also has the following flexible deployed form:

$$\underset{1\,2\,3\,4\,5\,6}{\exists\forall\exists\forall\exists\exists}\neg\rho x^1x^2\bigvee(\rho x^2x^5\bigwedge x^2\not\equiv x^5)\bigvee\rho x^2x^2$$
$$\bigvee\neg\rho x^3x^4\bigvee(\rho x^4x^6\bigwedge x^4\not\equiv x^6)\bigvee\rho x^4x^4$$

with extracted prefixes (1, 2, 5) and (3, 4, 6). We get a free thesis when 3 is replaced by 2 and 5 by 4.

As another example, note that for the formula considered in 6.3.1.4:

$$\underset{1\,2\,3\,4}{\exists\forall\exists\exists}\neg\rho x^1x^2\bigvee(\rho x^2x^3\bigwedge\rho x^3x^4),$$

when flexible deployed forms are used, order 3 (instead of 4) is sufficient to obtain the following explicit thesis:

$$\underset{1\,2\,3\,4\,5\,6\,7\,8\,9\,10\,11\,12}{\exists\forall\exists\forall\exists\forall\exists\exists\exists\exists\exists\exists}\quad\neg\rho x^1x^2\bigvee(\rho x^2x^7\bigwedge\rho x^7x^8)$$
$$\bigvee\neg\rho x^3x^4\bigvee(\rho x^4x^9\bigwedge\rho x^9x^{10})$$
$$\bigvee\neg\rho x^5x^6\bigvee(\rho x^6x^{11}\bigwedge\rho x^{11}x^{12}).$$

Take $h(3)=2$, $h(5)=h(7)=4$, $h(8)=6$.

6.6.2. *The completeness theorem remains valid if we require that there exist an explicit thesis which is a flexible deployed form in which the return number is* $u(i)=i-1$ *for every* $i\geqslant 2$.

$\rhd$ Define a partial order on the values assumed by the functions f_i in the flexible deployed form, by stipulating that, for arbitrary i and j, $f_i(x)\leqslant f_j(y)$ if $x\leqslant y$ and $f_i(x)=f_j(x)$. This partial order is a finite tree; all maximal chains which are restrictions of this tree are obtained by fixing i and considering all $f_i(x)$ for $x=1, 2,\ldots, n+p$. By an easy combinatorial lemma, we can enumerate the maximal chains which are restrictions of a finite tree, so that for each rank $i\geqslant 2$ the i-th chain reproduces the $(i-1)$-th chain up to a certain element, after which it contains new elements (not figuring in preceding chains). $\lhd$

6.6.3. We now describe an algorithm which transforms every flexible deployed form which is an explicit thesis into a strict deployed form which is also an explicit thesis (the result is due to J. P. Bénéjam).

(1) We divide the quantifier indices into sequences, which we call *periods*, each obtained by taking consecutive indices appearing in the same instance. In the example of 6.6.1, we have a flexible form with four periods: (1, 2) (3, 4), (5), (6).

(2) With each period U we associate a partial prefix U*, obtained by adjoining to U new indices which are stipulated to be immediate successors of those of U, and correspond to the ranks following those of U, up to $n+p$. In the preceding example, the four periods give the following partial prefixes: $(\underset{1}{\exists}\underset{2}{\forall}\underset{2'}{\exists})$, $(\underset{3}{\exists}\underset{4}{\forall}\underset{4'}{\exists})$, $(\underset{5}{\exists})$, $(\underset{6}{\exists})$.

(3) With each partial prefix U* obtained from the period U, we associate (in an obvious way) the renewal rank, equal to the rank of the first index of U; in the above example, these are the ranks 1, 1, 3, 3. The return number is obtained by returning to the period immediately preceding U and corresponding to the same instance in the flexible deployed form. If U is the first period, then the renewal rank is $n+1$ and return is unnecessary. Thus, in the above example, we go back from $\underset{5}{\exists}$ to the period (1, 2), therefore to (1, 2, 2′), and from $\underset{6}{\exists}$ to (3, 4), therefore to (3, 4, 4′). Thus we finally get the following strict deployed form, where it remains only to renumber the indices, in the order 1, 2, 2′, 3, 4, 4′, 5, 6:

$$\underset{1}{\exists}\underset{2}{\forall}\underset{2'}{\exists} \quad \neg \rho x^1x^2 \vee (\rho x^2x^{2'} \wedge x^2 \not\equiv x^{2'}) \vee \rho x^2x^2$$
$$\underset{3}{\exists}\underset{4}{\forall}\underset{4'}{\exists} \vee \neg \rho x^3x^4 \vee (\rho x^4x^{4'} \wedge x^4 \not\equiv x^{4'}) \vee \rho x^4x^4$$
$$\underset{5}{\exists} \vee \neg \rho x^1x^2 \vee (\rho x^2x^5 \wedge x^2 \not\equiv x^5) \vee \rho x^2x^2$$
$$\underset{6}{\exists} \vee \neg \rho x^3x^4 \vee (\rho x^4x^6 \wedge x^4 \not\equiv x^6) \vee \rho x^4x^4 .$$

A free thesis is obtained here by performing the same substitutions as for the original flexible deployed form (3 by 2 and 5 by 4).

6.7. Interpolation theorem [CRA, 1957; LYN, 1959]

An index i is said to be *inactive in an operator* $\mathscr{P}$ or *in a formula representing*

$\mathscr{P}$ if, for any multirelation M assignable to $\mathscr{P}$, the index i is inactive in the relation $\mathscr{P}(M)$. Recall that this means that

$$\mathscr{P}(M)(x_1, \ldots, x_i, \ldots, x_n) = \mathscr{P}(M)(x_1, \ldots, x', \ldots, x_n)$$

for any value x' from the base (see 3.6.1).

6.7.1. *Let* A, B *be two free formulas of the same predicarity,* U, V *two sequences of quantifiers such that* $UA \vdash VB$. *Assume moreover that the indices of* U *and those of* V *are consecutive, the first index of* V *following the last index of* U, *the indices of* U *are inactive in* B *and the indices of* V *are inactive in* A. *Then there exist two free formulas* A′, B′ *and a sequence* W *of quantifiers such that*

(1) $A' \vdash B'$;

(2) $UA \vdash WA' \vdash WB' \vdash VB$;

(3) *every predicate inactive in* A *is inactive in* A′; *any predicate inactive in* B *is inactive in* B′.

▷ Let U^* denote the dual prefix, i.e., the prefix obtained from U by interchanging ∀ and ∃. Then $U^*V(A \Rightarrow B)$ is a canonical prenex thesis. By the completeness theorem, there exists an explicit thesis which is a deployed form of the above. That is to say, there exist instances $A_1, \ldots, A_h$ of A and $B_1, \ldots, B_h$ of B, sequences of quantifiers $U_1, \ldots, U_h$ with duals $U_1^*, \ldots, U_h^*$, and sequences of quantifiers $V_1, \ldots, V_h$, such that the following formula is an explicit thesis:

$$U_1^* V_1 \ldots U_h^* V_h (A_1 \wedge \cdots \wedge A_h) \Rightarrow (B_1 \vee \cdots \vee B_h).$$

Let A_i' and B_i' $(i = 1, \ldots, h)$ denote bi-instances obtained from A_i and B_i when the ∃-indices are replaced by smaller integers.

Set

$$A' = A_1' \wedge \cdots \wedge A_h', \qquad B' = B_1' \vee \cdots \vee B_h'.$$

The formula $A' \Rightarrow B'$ is a free thesis; therefore $A' \vdash B'$, which is condition (1). Moreover, every predicate inactive in A remains inactive in its instances A_i and its bi-instances A_i' $(i = 1, \ldots, h)$, therefore also in A′. The same holds for B and B′. Thus condition (3) holds. Now set $W = U_1 V_1 \ldots U_h V_h$. The condition $WA' \vdash WB'$ follows from (1), and hence it remains to prove that $UA \vdash WA'$, or what is the same, the dual condition $WB' \vdash VB$.

Note first that $V_1 \ldots V_h (B_1 \vee \cdots \vee B_h)$ is a deployed form of VB, and is therefore equideducible from it (see 6.2). Replacing certain $\exists$-indices of W by smaller integers, we obtain $B' = B_1' \vee \cdots \vee B_h'$ from $B_1 \vee \cdots \vee B_h$. Thus $WB' \vdash W(B_1 \vee \cdots \vee B_h)$ (3.8.6). Since the indices of U are inactive in B, the indices of $U_1, \ldots, U_h$ are inactive in $B_1 \vee \cdots \vee B_h$, and hence

$$W(B_1 \vee \cdots \vee B_h) \dashv\vdash V_1 \ldots V_h(B_1 \vee \cdots \vee B_h),$$

and finally

$$WB' \vdash V_1 \ldots V_h(B_1 \vee \cdots \vee B_h) \dashv\vdash VB. \ \triangleleft$$

6.7.2. THEOREM. *Let* A, B *be two free formulas of the same predicarity, and* U, V *two sequences of quantifiers such that* $UA \vdash VB$. *Then there exists a free formula* C *of the same predicarity, in which the only active predicates are those active both in* A *and in* B, *and a prefix* W *such that* $UA \vdash WC \vdash VB$.

$\triangleright$ We may assume without loss of generality that the indices of U and of V are consecutive, the first index of V follows the last index of U, the indices of U are inactive in B and those of V inactive in A. Otherwise we need only apply to UA and VB two suitable bijective functions (equal to the identity on free indices). Now consider the free formulas A′, B′ and prefix W which exist according to 6.7.1. By 4.5.5, there exists a free formula C such that $A' \vdash C \vdash B'$, where the only active predicates in C are those active both in A′ and in B′, therefore both in A and in B. Moreover, we have $WA' \vdash WC \vdash WB'$, and this, combined with (2) of 6.7.1, implies $UA \vdash WC \vdash VB$. $\triangleleft$

As an immediate consequence, *let* $\mathscr{A}, \mathscr{B}$ *two logical classes with* $\mathscr{A} \subseteq \mathscr{B}$; *then there exists a logical class* $\mathscr{C}$ *such that* $\mathscr{A} \subseteq \mathscr{C} \subseteq \mathscr{B}$, *the only active predicates in* $\mathscr{C}$ *being those active both in* $\mathscr{A}$ *and in* $\mathscr{B}$.

EXERCISES

1

Consider a canonical prenex formula P = UA with prefix U and free part A. With each $\forall$-index i, we associate a function t_i defined for all sequences of positive integers indexed by the $\exists$-indices $< i$ of U, with positive integer values.

(1) Let M be a multirelation assignable to P and defined on a set E of positive integers. Show that P(M) = + if and only if, for any functions t_i under which E is closed, there exist elements of E assignable to the $\exists$-indices which, when the corresponding values of the t_i's are assigned to the $\forall$-indices, give A(M) the value +. For example, the quaternary relation R assignable to the predicate ρ gives $\underset{1234}{\exists\forall\exists\forall}\ \rho x^1x^2x^3x^4$ the value + if and only if, for any functions t_2 and t_4, there exist a_1 and a_3 such that $R(a_1, t_2(a_1), a_3, t_4(a_1, a_3)) = +$.

(2) A function, defined on the positive integers with positive integer values, is said to be *strictly majorant* if its value is always strictly greater than its maximum argument. By the preceding part of the exercise, if P(M) = +, then, for any strictly majorant functions t_i, there exist elements of the base assignable to the $\exists$-indices which, when the corresponding values of the t_i's are assigned to the $\forall$-indices, give A(M) the value +. The converse is false, as can be seen from the formula $\underset{12}{\exists\forall}\ \rho x^1x^2$, by assigning to the predicate ρ the relation R such that $R(a, b)$ if and only if $a < b$ (a and b positive integers): the formula assumes the value −, whereas, for strictly majorant t and any element a, we have $R(a, t(a)) = +$.

Nevertheless, show that if the condition holds for all M (more precisely, if, for every M over a set E of positive integers and any strictly majorant functions t under which E is closed, there exist elements of E which, together with the values of t, give A(M) the value +), then P is a thesis. Indirectly: assuming that P is not a thesis, construct a completed deployment as in 6.5, and use it to construct strictly majorant functions which always assign the value − to A(M).

With regard to the preceding arguments, we remark that the original idea of the deployed form dates back to [SKO, 1920]; it involved replacing the $\forall$-indices by a function symbol followed by certain smaller $\exists$-indices. A free thesis was derived as in 6.3, but without the intermediate step of an explicit thesis. For example, in the thesis

$$\underset{1234}{\exists\forall\exists\exists} \neg\rho x^1x^2 \vee (\rho x^2x^3 \wedge \rho x^3x^4)$$

we must replace 2 by $t(1)$, and then proceed to any of the free theses already discussed above, such as that of 6.6.1, which here has the form

$$\begin{aligned}&\neg\rho x^1x^{t1} \vee (\rho x^{t1}x^7 \wedge \rho x^7x^8) \vee \neg\rho x^3x^{t3}\\ &\quad \vee (\rho x^{t3}x^9 \wedge \rho x^9x^{10}) \vee \neg\rho x^5x^{t5} \vee (\rho x^{t5}x^{11} \wedge \rho x^{11}x^{12}),\end{aligned}$$

in which we now replace 3 by $t(1)$, 5 and 7 by $t(3)$, 8 by $t(5)$. We can also replace $t(3)$ by $t(t(1))$ and $t(5)$ by $t(t(t(1)))$, but this is superfluous.

2

Consider the following two logical formulas:

$$\underset{123}{\forall\forall\exists}(\rho x^1x^1 \wedge \rho x^1x^2) \vee \neg\rho x^1x^3,$$

$$\underset{1234}{\forall\exists\forall\exists}(\rho x^1x^2 \wedge \rho x^3x^3 \wedge \rho x^2x^4) \vee \neg\rho x^4x^3.$$

(1) Show that they are theses, and construct in each case a deployed form which is an explicit thesis. For each formula, specify the minimum order of such a deployed form.

(2) Show that in neither case can we obtain an explicit thesis if we assume that each instance has renewal rank one, i.e., if the entire sequence of quantifiers is repeated with new values for each instance. The same holds for flexible deployed forms.

(3) For the second formula, show that we cannot obtain an explicit thesis if we assume that the return number is one for each instance, i.e., if we always come back to the first instance (this result is due to J. P. Bénéjam). Show that this is no longer true for flexible deployed forms (see 6.6): there exists a flexible deployed form of order 3, with returns to the first instance, which is an explicit thesis.

3

Consider the free formulas:

$$\begin{aligned} A &= \neg\rho x^1x^2 \bigvee (\rho x^2x^3 \bigwedge \rho x^3x^3), \\ B &= \neg\sigma x^1x^2 \bigvee (\sigma x^1x^3 \bigwedge \sigma x^3x^3), \\ C &= \neg\rho x^1x^2 \bigvee B. \end{aligned}$$

(1) Show that each of the formulas A, B, C, $A \bigwedge C$ becomes a thesis when prefixed with $\underset{1\,2\,3}{\exists\exists\forall}$. Find for each an explicit thesis which is a deployed form of order 3 for the first three, of order 7 for the fourth.

(2) Show that $\underset{1\,2\,3}{\exists\exists\forall}(A \bigwedge B)$ is not a thesis: define a multirelation on two elements which assigns the formula the value $-$.

4

Consider the three theses of Chapter 5, Exercise 4.2:

$$\begin{aligned} &\underset{2\,3\,4}{\exists\forall\exists}(\rho x^1x^2 \bigwedge \rho x^2x^2 \bigwedge \rho x^3x^3) \bigvee \neg\rho x^4x^3, \\ &\underset{2\,3\,4}{\exists\forall\exists}(\rho x^1x^2 \bigwedge \rho x^2x^4 \bigwedge \rho x^3x^3) \bigvee \rho x^4x^3, \\ &\underset{1\,2\,3}{\exists\forall\exists}(\rho x^1x^1 \bigwedge \rho x^2x^2 \bigwedge \rho x^1x^3) \bigvee \neg\rho x^3x^2. \end{aligned}$$

(1) For the first formula, find a deployed form of order 3, with two returns to the first instance, which is an explicit thesis.

(2) For the second and third formulas, find a deployed form of order 3, with a return to the second instance, which is an explicit thesis. Show that these formulas cannot be identified as theses by a deployed form of arbitrary order in which all the returns are to the first instance (see [BEN, 1969]).

5

Consider the following thesis, which contains two predicates ρ, σ:

$$\underset{1\,2\,3}{\exists\forall\exists}(\rho x^1x^1 \bigwedge \rho x^2x^2 \bigwedge \rho x^1x^3 \bigwedge \sigma x^1x^1 \bigwedge \sigma x^1x^2) \bigvee \neg\rho x^3x^2 \bigvee \neg\sigma x^2x^3.$$

Find a deployed form of order 4 which is an explicit thesis. Show that, even for flexible deployed forms, there must be at least one return number $\geqslant 2$ (see [BEN, 1969]).

Find a deployed form of order 5 for the following thesis (this example is also due to Bénéjam):

$$\underset{2\,3\,4}{\exists\forall\exists}(\rho x^1x^2 \bigwedge \neg\rho x^4x^1 \bigwedge \neg\rho x^2x^4) \bigvee (\rho x^3x^4 \bigwedge \neg\rho x^2x^4) \bigvee \neg\rho x^2x^2 \bigvee \rho x^3x^2.$$

Problem. Does there exist a thesis for each integer h such that every deployed form which is an explicit thesis contains returns to h distinct instances? Does a thesis of this type exists if we restrict ourselves to unary predicates? What is the situation as regards a thesis with one binary predicate (in the above example, $h=2$, but there are two binary predicates)?

6

Let P be a nonprenex thesis, in which the connections are $\bigwedge$, $\bigvee$ and occurrences of $\neg$ immediately preceding the free operators. Suppose that the indices of the different occurrences of quantifiers have been made distinct. We now ask: what is the best strategy for transforming P into a prenex thesis? To be precise, how should we arrange the quantifiers at the head of the formula, so that the resulting prenex formula will have a deployed form of minimum order which is an explicit thesis?

(1) We systematically give priority to the quantifier $\forall$ over the quantifier $\exists$, whenever the two types appear among those that can be brought forward immediately. When the quantifiers to be brought forward are either all $\forall$ or all $\exists$, priority is given to that dominating the shortest subformula. Note that this is a rather bad strategy: starting from the formula

$$\underset{1}{\exists}\underset{2}{\exists}\underset{3}{\forall}(x^1 \not\equiv x^3 \bigvee x^2 \not\equiv x^3 \bigvee \rho x^3) \bigvee \underset{4}{\exists}(\neg \rho x^4 \bigwedge \underset{5}{\forall} x^4 \equiv x^5) \bigvee \underset{6}{\exists}\underset{7}{\exists} x^6 \not\equiv x^7,$$

this strategy instructs us to bring forward the quantifiers in the order 6, 7, 4, 5, 1, 2, 3, and the result is not an explicit thesis. On the other hand, the order 1, 2, 3, 4, 5, 6, 7 gives an explicit thesis.

(2) Modify the preceding strategy, stipulating that, when the quantifiers to be brought forward are all $\forall$ or all $\exists$, priority is given to that which dominates the longest subformula. By a slight modification of the above formula (lengthening the subformulas dominated by $\underset{4}{\exists}$ and $\underset{6}{\exists}$), show that we again get a bad strategy, again giving the order 6, 7, 4, 5, 1, 2, 3.

(3) Give priority to the quantifiers $\forall$ not dominated by a $\exists$; then take the quantifiers $\forall$ dominated by a single quantifier $\exists$, followed by quantifiers $\forall$ dominated by two $\exists$'s, etc., and finally quantifiers $\exists$ not dominating any $\forall$. This gives the order 4, 5, 1, 2, 3, 6, 7, which yields an explicit thesis. Find an example showing that this strategy is nevertheless not good.

CHAPTER 7

INTERPRETABILITY OF RELATIONS

7.1. Interpretability; general considerations

Let M be a multirelation. A relation S over the same base is said to be *interpretable* by M if there exists a logical formula P such that $P(M)=S$ (or, equivalently, a logical operator $\mathscr{P}$ such that $\mathscr{P}(M)=S$). A multirelation N over the same basis as M is *interpretable* by M if each relation of N is interpretable by M. This concept is a generalization of free interpretability (see 4.2 and 4.3.5).

Examples. Let the base be the set of natural numbers; let A_0 be the unary relation $A_0(x)=+$ when $x=0$ and $-$ when $x\neq 0$. Let $I(x,y)=+$ when $x\leqslant y$, and $C(x,y)=+$ when y is the successor of x. Then A_0 is interpretable by I, by means of the formula $\forall \iota xy$ (where ι is a binary predicate, therefore replaceable by I). The relation C is interpretable by I by means of the formula

$$\iota xy \wedge x\not\equiv y \wedge \forall_z (\iota zx \vee \iota yz).$$

Finally, A_0 is interpretable by E, the ternary relation "z is between x and y", by means of the formula

$$\neg \exists_{yz} \varepsilon yzx \wedge y\not\equiv x \wedge z\not\equiv x,$$

where ε is a ternary predicate, therefore replaceable by E.

7.1.1. *Interpretability is a pre-ordering.*

Interpretability is obviously reflexive: any n-ary relation R is interpretable by itself by means of the formula $\rho x^1 \ldots x^n$. We must show that interpretability is transitive. Let M, M′, M″ be three multirelations over the same base. Suppose that each relation of M″ has the form $\mathscr{P}(M')$ and each relation $R_i\,(i=1,\ldots,h)$ of $M'=R'_1,\ldots,R'_h$ has the form $R'_i=\mathscr{Q}_i(M)$, where $\mathscr{P}$ and the $\mathscr{Q}_i$ are logical operators. Then each relation of M″ has the form $\mathscr{P}(\mathscr{Q}_1(M),\ldots,\mathscr{Q}_h(M))$. But, by 5.2.2, the composite operator $\mathscr{P}(\mathscr{Q}_1,\ldots,\mathscr{Q}_h)$ is logical.

7.1.2. *If* N *is interpretable by* M, *then any automorphism of* M *is an automorphism of* N.

In fact, by 5.2, every automorphism of M is an automorphism of $\mathscr{P}$(M), for any logical operator $\mathscr{P}$.

For example, it follows that the chain I of natural numbers is not interpretable by A_0 (the relation equal to + for the single element 0), since the permutation which interchanges 1 and 2 and leaves all other numbers unchanged is an automorphism of A_0 but not of I.

The product P of natural numbers, defined by $P(x, y, z) = +$ when $x.y = z$, admits several automorphisms; among these is the permutation which leaves 0, 1 unchanged, interchanges 2 and 3, and, in general, interchanges any two numbers whose prime factorizations are $2^a.3^b.5^c\ldots$ and $2^b.3^a.5^c\ldots$. Thus neither sum, chain, nor successor relation over the natural numbers are interpretable by P (see [PAD, 1902]).

It has been proved that the product of natural numbers is not interpretable by the sum ([PRE, 1929]; see Volume 2, Section 2.5.4).

7.1.3. *If* M *and* N *have the same finite base and any automorphism of* M *is an automorphism of* N, *then* N *is interpretable by* M.

▷ Let $a, \ldots, b$ be a finite sequence of base elements; denote the other elements by a_i $(i = 1, 2, \ldots, r)$. For any relation R of M, consider the conjunction of identifiers and negations of identifiers which state the various equalities and inequalities holding among the elements $a, \ldots, b$ and $a_1, \ldots, a_r$, and of the rank-changers which state the values assumed by R. If the indices $1, \ldots, r$ are then associated with the quantifier $\exists$, we get a formula $P_{a, \ldots, b}$ which is true for the system (M; $a, \ldots, b$) and only for any system consisting of M and a sequence $a', \ldots, b'$ derived from $a, \ldots, b$ by an automorphism of M. For any (n-ary) relation S of N and any n-tuple $a, \ldots, b$, consider the formula

$$\forall_{x \ldots y} P_{a \ldots b}(x, \ldots, y) \Rightarrow \sigma x \ldots y$$

if $S(a, \ldots, b) = +$, and the formula obtained from the latter by negating σ if $S(a, \ldots, b) = -$. The conjunction of all these formulas interprets N by M. ◁

Note that if the base is infinite and every automorphism of M is an automorphism of N, it does not necessarily follow that N is interpretable by M. For example, the chain I and the sum relation, S $(x, y, z) = +$ when

$x+y=z$ (over the natural numbers), both admit only the identity automorphism. I is indeed interpretable by S, by means of the formula $\underset{z}{\exists}\sigma xzy$, but S is not interpretable by I (see Volume 2, Section 1.7.1).

7.1.4. *Addition over the natural numbers is interpretable by* (P, C), *therefore also by* (P, I), where C is the successor relation and I the chain. This can be seen from the fact that $x+y=z$ is equivalent to the disjunction of the following two conditions: (a) $x=y=z=0$; (b) $z\neq 0$ and $(xz+1)(yz+1)=$ $=(xy+1)z^2+1$ (see [ROB, 1949]).

Let D be the divisibility relation over the natural numbers: $D(x, y)=+$ if x divides y (or $y=0$). Note that the singleton zero (the unary relation equal to + for the single element 0) and the singleton one are interpretable by D, for 0 is the only integer divisible by all integers and 1 the only integer which divides all integers.

The relation E *defined as* $E(x, y)=+$ *if* x *and* y *are relatively prime integers is interpretable in* D, for E is equivalent to the statement that 1 is the only integer dividing both x and y. *The ternary relation* M *defined by* $M(x, y, z)=z$ *when* z *is the lowest common multiple of* x *and* y (or $x=z=0$ or $y=z=0$) *is interpretable in* D: the condition simply means that every common multiple of x and y is a multiple of z and conversely.

7.1.5. *The product relation over the natural numbers, and therefore also the sum, is interpretable in* (C, D), *where* C *is the successor relation and* D *divisibility* [J. ROB, 1949].

▷ By 7.1.4, it will suffice to prove that the product is interpretable in (C, D, E, M) where E is the relation "relatively prime" and M the relation "lowest common multiple".

Let a, b be two integers whose product ab is at least 2. Let x be a nonzero integer prime to a, y a nonzero integer prime to b, and m a common divisor of $ax+1$ and $by+1$. In other words, $ax=-1$ and $by=-1$ (mod m). Then $abxy=1$ (mod m), and hence the integer $u=abxy-1$ is not zero and is a multiple of m. Moreover, the product ax is the same as the lowest common multiple of a and x, and similarly for b and y.

Now let a, b, c be three nonzero integers, $c\geqslant 2$. Let m be an arbitrarily large integer prime to a and b. By Bezout's equality, there exists an integer x such that $ax=-1$ (mod m). Since $m\geqslant 2$, this integer is not zero. Moreover, it is prime to m by the preceding equality. It follows that, for any finite set of integers, in particular, for a and c, there exists i such that $x+im$ is prime to all integers of the set, in our case a and c. Moreover, $a(x+im)=-1 \pmod{m}$. Finally, there exists a nonzero integer x, prime to a and c, such that $ax=-1 \pmod{m}$. Similarly, there exists a nonzero integer y, prime to b, c, x, such that $by=-1 \pmod{m}$. Now $u=cxy-1$ is not zero; moreover, if we suppose u to be a multiple of m it follows that $cxy=1 \pmod{m}$ and so $cxy=abxy \pmod{m}$; thus $c=ab \pmod{m}$ since m is prime to x and y. But since this is true for arbitrarily large m, it follows that $ab=c$.

It follows from the preceding two paragraphs that the equality $ab=c$ is equivalent to the disjunction of the following four conditions: (1) $a=c=0$; (2) $b=c=0$, (3) $a=b=c=1$; (4) a, b are not 0, c is neither 0 nor 1, and, for any x prime to a, c, any y prime to b, c, x, and m dividing the two integers (l.c.m. a, x)+1, (l.c.m. b, y)+1, the integer $u=$(l.c.m. c, x, y)-1 is a multiple of m. ◁

7.2. Interpretability by sum and product of natural numbers; arithmetical relation

Relations over the natural numbers which are interpretable by (S, P), where S is the ternary relation "sum" and P the relation "product", are known as *arithmetical relations* (they can be defined in a more general manner, requiring no reference to the two special relations S and P, but to recursive relations).

A set of integers is said to be *arithmetical* if the unary relation of membership in the set is arithmetical. Intersection, union, and complement of arithmetical sets are arithmetical sets. The image of an arithmetical set A under an arithmetical relation R (the set of all y such that there exists $x \in A$ with $R(x, y)=+$) is an arithmetical set; this remains true for the inverse image. More generally, starting from arithmetical relations and applying a logical operator leads to an arithmetical relation.

7.2.1. Let us prove the following elementary properties of finite sequences of integers:

(1) *If n and v are positive integers with v divisible by* $1, 2, \ldots, n$, *then the integers $b_i=v(i+1)+1$ $(i=0, 1, \ldots, n)$ are relatively prime in pairs.* In fact, any prime number dividing both $v(i+1)+1$ and $v(j+1)+1$ $(0 \leqslant i < j \leqslant n)$

must also divide $(j-i)\,v$. It therefore divides either $j-i$ or v. Hence it must divide v. But then our prime number divides $v(i+1)$, which is impossible since then it would have to divide 1.

(2) *Let* $a_0, a_1, \ldots, a_n$ *be a sequence of natural numbers. There exists an integer* m *such that, for any* $v \geqslant m$ *divisible by* $1, 2, \ldots, n$, *there exists* u *with the property: "*a_i *= remainder upon division of* u *by* $(i+1)\,v+1$*" for* $i=0, 1, \ldots, n$. The statement is true for $n=0$: it suffices to take $v \geqslant a_0$ and $u=a_0$. Suppose the statement true for $n-1$, and add an integer a_n to the given sequence a_i $(i=0, 1, \ldots, n-1)$. Define a new m as the maximum of the old m and a_n; now let v be divisible by $1, 2, \ldots, n$ and $v \geqslant m$ and $\geqslant a_n$. We know that there exists u such that a_i = remainder upon division of u by $(i+1)\,v+1$ for $i=0, 1, \ldots, n-1$. Set $w=(v+1)(2v+1)\ldots(nv+1)$. Nothing is changed in the above remainders if we replace u by u + a multiple of w. Moreover, $v+1, 2v+1, \ldots, nv+1$ are all prime to $(n+1)\,v+1$ by (1). Hence their product w is also prime to $(n+1)\,v+1$. Thus the remainders upon division of $u, u+w, u+2w, \ldots, u+(n+1)\,vw$ by $(n+1)\,v+1$ are distinct: these are precisely the numbers $0, 1, \ldots, (n+1)\,v$, in another order. Since one of them is $a_n \leqslant v \leqslant (n+1)\,v$, this gives us a new integer u.

We shall need the following immediate corollary (2′) of (2):

(2′) *If* $a_0, a_1, \ldots, a_n$ *is a sequence of natural numbers, there exist* u *and* v *such that "*a_i *= remainder upon division of* u *by* $(i+1)\,v+1$*" for* $i=0, 1, \ldots, n$.

For another proof, see [MEN, 1964], p. 135.

7.2.2. *We now show that the following relations are arithmetical:* the *factorial*, i.e., the relation defined as $\mathrm{F}(x, y)=+$ when $y=x!$, and the *exponential*:

$$\mathrm{Exp}(x, y, z)=+ \quad \text{when} \quad z=x^y$$

([GOD, 1930]; see also [KLE, 1952]).

In general, *all relations defined by recursion from addition and multiplication are arithmetical.* Using the exponential, we can show that the "Fermat relation" is arithmetical: this is the relation defined as + for n if and only if there are three positive integers x, y, z such that $x^n+y^n=z^n$.

Proof of 7.2.2. ▷ In what follows the ternary predicates σ and $\bar{\omega}$ will be replaceable by the relations S and P. The unary relation A_0 equal to + for 0 alone is interpretable by S, by means of the formula σxxx, and is therefore arithmetical. The relation A_1 defined as + for 1 alone is interpretable by S, by means of the formula

$$\underset{y}{\exists}\sigma yyy \wedge x \not\equiv y \wedge \underset{z}{\forall}\left(z \equiv y \vee \underset{t}{\exists}\sigma xtz\right).$$

Similarly, we see that the relation A_n equal to + for n alone is interpretable by S.

The chain I of natural numbers is interpretable by S, as we have already seen, by means of the formula $\underset{z}{\exists}\sigma xzy$. The strict chain I*, equal to + when $x<y$, is interpretable by means of $\underset{z}{\exists}\sigma xzy \wedge x \not\equiv y$. The successor relation C, defined as + when $y=x+1$, is interpretable by (S, A_1) via the formula $\underset{z}{\exists}\sigma xzy \wedge \alpha_1 z$, where α_1 is replaceable by A_1. Thus C is interpretable by S alone.

The quaternary relation equal to + when division of x by y $(y \neq 0)$ yields quotient z and remainder t is interpretable by (S, P, A_0, I*), by means of the formula

$$\neg\alpha_0 y \wedge \underset{u}{\exists}(\bar{\omega} yzu \wedge \sigma utx) \wedge \iota^* ty,$$

where α_0 is replaceable by A_0 and ι^* by I*; thus it is arithmetical. Consequently, the ternary relation Q equal to + when z is the quotient upon division of x by $y \neq 0$, and the relation R equal to + when z is the remainder upon division of x by $y \neq 0$, are both arithmetical. The quaternary relation T equal to + when t is the remainder upon division of x by $y(z+1)+1$ is arithmetical, since it is interpretable in P, C, R by the formula

$$\underset{z'uu'}{\exists}\gamma zz' \wedge \bar{\omega} yz'u \wedge \gamma uu' \wedge \rho xu't,$$

where γ is replaceable by C and ρ by R.

To see that the factorial relation is arithmetical, we observe that $y=x!$ is equivalent, by 7.2.1 (2′), to the following condition:

"There exist two integers u and v such that $1=$ remainder upon division of u by $v+1$, $y=$ remainder upon division of u by $(x+1)\,v+1$, and moreover, for $i \leqslant x-1$ the remainder upon division of u by $(i+1)\,v+1$, when multiplied by $i+1$, gives the remainder upon division of u by $(i+2)\,v+1$."

This condition shows that the factorial is interpretable in (A_0, A_1, T, I*, C, P) by the following formula (in which τ is replaceable by the relation T):

$$\underset{uv}{\exists}\left[\left(\underset{tt'}{\exists}\alpha_0 t \wedge \alpha_1 t' \wedge \tau uvtt'\right)\wedge \tau uvxy\right.$$
$$\left.\wedge\left(\underset{ii'jj'}{\forall}\iota^* ix \wedge \gamma ii' \wedge \tau uvij \wedge \tau uvi'j' \Rightarrow \bar{\omega} ji'j'\right)\right].$$

For example, to compute $4!=24$, we can take $u=976431$ and $v=6$; the integers $(i+1)v+1$ from $i=0$ to 4 are 7, 13, 19, 25, 31, which give the desired remainders:

$$\begin{aligned} 976431 &= (\text{multiple of } 7)+1 \\ &= (\text{multiple of } 13)+1 \\ &= (\text{multiple of } 19)+2 \\ &= (\text{multiple of } 25)+6 \\ &= (\text{multiple of } 31)+24. \end{aligned}$$

To see that the exponential is an arithmetical relation, we observe that $z=x^y$ is equivalent to the following condition:

"There exist u and v such that $1=$remainder upon division of u by $v+1$, $z=$remainder upon division of u by $(y+1)v+1$; moreover, for $i \leqslant y-1$ the remainder upon division of u by $(i+1)v+1$, when multiplied by x, gives the remainder upon division of u by $(i+2)v+1$."

This condition shows that the exponential is indeed interpretable by the multirelation (A_0, A_1, T, I*, C, P), by the following formula:

$$\underset{uv}{\exists}\left[\left(\underset{tt'}{\exists}\alpha_0 t \wedge \alpha_1 t' \wedge \tau uvtt'\right)\wedge \tau uvyz\right.$$
$$\left.\wedge\left(\underset{ii'jj'}{\forall}\iota^* iy \wedge \gamma ii' \wedge \tau uvij \wedge \tau uvi'j' \Rightarrow \bar{\omega} xjj'\right)\right]. \lhd$$

7.2.3. To obtain a nonarithmetical set, we can use the fact that the family of arithmetical sets must be denumerable, and apply the diagonal method. Without explicit appeal to logic, one can only indicate examples which are not known to be arithmetical. Thus, the set of integers n such that $2^{\aleph_n}=\aleph_{n+1}$ is such a set; recall that $\aleph_0$ is the denumerable cardinal, and $\aleph_{n+1}$ is the cardinal of the set of ordinals of cardinal $\aleph_n$ (this example is due to G. Kreisel). We are obviously assuming that the set theory in which logic is embedded (see beginning of Chapters 1 and 3) does not contain the generalized continuum axiom, for otherwise the set in question would simply be the set of all natural numbers. If we assume the axiom of choice and the generalized continuum axiom, no example is known which is definable without appeal to logic and which is not known to be arithmetical.

7.3. Interpretability of the Set of Natural Numbers by the Sum and Product of Integers, of Rationals, of Reals

7.3.1. Consider the relations S and P (sum and product) over the integers. The *unary relation* N such that $N(x) = +$ when x is a natural number (nonnegative integer) *is interpretable by* (S, P). It suffices to use Lagrange's theorem on the expression of any natural number as a sum of four squares to see that $N(x) = +$ if and only if there exist integers $s, t, u, v, s', t', u', v', y, z$ such that $P(s, s, s')$, $P(t, t, t')$, $P(u, u, u')$, $P(v, v, v')$, $S(s', t', y)$, $S(y, u', z)$ and $S(z, v', x)$ are all equal to $+$. Consequently, the usual ordering $\leqslant$ of the integers is interpretable by (S, P).

The relation N is not interpretable by S alone, for the permutation of the base which changes the sign of every nonzero integer preserves S but not N. Similarly, N is not interpretable by P alone, for the permutation which changes the sign of every integer whose prime factorization involves an odd power of 2 (e.g., 2, 6, 8, 10, 14,...) preserves P but not N.

Problem. It is not known whether N is interpretable by the sum S and the singleton relation 1.

7.3.2. Consider the relations S and P over the rationals. One can show that every unary singleton relation (equal to + for a single rational number) is interpretable by (S, P): for 0 is the only element such that $u + u = u$, and 1 is the only nonzero element such that $v.v = v$; thus we can define the singletons $2 = 1 + 1$, $3 = 2 + 1$, etc., and finally the singleton for any rational x, the quotient of integers p and q, by the condition $q.x = p$.

The unary relation A equal to + for the positive rationals is interpretable by (S, P), again via Lagrange's theorem. In fact, every positive rational has the form $p/q = pq/q^2$ with positive integers p, q. The product $p.q$ is a sum of four squares, say t^2, u^2, v^2, w^2, so that p/q is the sum of the squares of $t/q, u/q, v/q, w/q$. In other words, the positive rationals, and they alone, are all sums of four square rationals. It follows that the usual ordering $\leqslant$ of the rational numbers is interpretable by (S, P).

7.3.3. *The unary relation over the rationals, defined as + for the integers, is interpretable in* (S, P) (sum and product of rationals) [J. ROB, 1949]

▷ The proof requires the following two arithmetical lemmas, which we assume without proof: (1) If p is a prime integer congruent to -1 modulo 4, there exist three rationals r, s, t such that $px^2+pt^2+2=r^2+s^2$ if and only if the denominator of the reduced fraction equal to x is not divisible by 2 or p. (2) If p is a prime integer congruent to $+1$ modulo 4, and q is an odd prime integer which is a quadratic non-residue of p, then there exist three rationals r, s, t such that $pqx^2+pt^2+2=r^2+qs^2$ if and only if the denominator of the reduced fraction equal to x is not divisible by p or q (see Volume 2).

Now consider the following formula $\mathrm{P}(u, v, x)$:

$$\exists_{r,s,t}\; uvx^2+vt^2+2=r^2+us^2$$

and interpret the set of integers by the following formula:

$$\forall_{u,v}\,[\mathrm{P}(u, v, 0) \wedge \forall_{y}(\mathrm{P}(u, v, y) \Rightarrow \mathrm{P}(u, v, y+1))] \Rightarrow \mathrm{P}(u, v, x).$$

This formula is satisfied whenever x is an integer. Indeed, if u, v are rational numbers, then either $\mathrm{P}(u, v, 0) = -$ and the formula is satisfied, or $\forall_{y}(\mathrm{P}(u, v, y) \Rightarrow \mathrm{P}(u, v, y+1)) = -$ and the formula is again satisfied, or, finally, both the above formulas have the value + and then $\mathrm{P}(u, v, x) = +$ for every positive integer x, and also for every negative integer, since x appears in the above equality only as a square.

Conversely, let x be a rational number satisfying the formula. In particular, x will satisfy the particularized formula with 1 substituted for u, a prime number p congruent to -1 modulo 4 for v. Now $\mathrm{P}(1, p, 0)$ takes the value +, as shown by setting $r=s=1$ and $t=0$. Moreover, the particularized formula $\forall_{y}(\mathrm{P}(1, p, y) \Rightarrow \mathrm{P}(1, p, y+1))$ takes the value +, since the denominator of the reduced fraction equal to y and that corresponding to $y+1$ are the same (apply lemma (1)). Thus, if a rational number x satisfies the formula, then $\mathrm{P}(1, p, x)$ takes the value +; by lemma (1), we see, varying p, that the denominator of the reduced fraction has neither 2 nor any integer congruent to -1 modulo 4 as a prime factor.

Moreover, if p is an odd prime, there exists an odd prime q which is a quadratic non-residue of p. For let q' be a non-residue of p. Either q' is odd, or $q'+p$ is odd and a non-residue; it has at least one non-residue prime factor q, which must be odd. If x satisfies the formula, then it satisfies the particularized formula with q, p substituted for u, v, where p is a prime congruent to $+1$ modulo 4 and q is an odd prime which is a non-residue of p. Then $\mathrm{P}(q, p, 0)$ takes the value $+$, since the denominator of the reduced fraction 0/1 is 1 (apply lemma (2)). Moreover,

$$\forall_y (\mathrm{P}(q, p, y) \Rightarrow \mathrm{P}(q, p, y+1))$$

assumes the value $+$, since the denominator of the reduced fraction is the same for y and $y+1$ (again by lemma (2)). Thus $\mathrm{P}(q, p, x)$ takes the value $+$; by lemma (2), we see, varying p, that the denominator of the reduced fraction has no prime factor congruent to $+1$ modulo 4. It now follows that the only admissible "prime" factor of the denominator is 1, and so the rational number x is an integer. $\lhd$

It follows from 7.3.2 and 7.3.3 that the unary relation over the rationals equal to $+$ for the natural numbers is interpretable by (S, P).

7.3.4. Let S and P be the sum and product over the real numbers. One proves as in 7.3.2 that the singleton of 0, the singleton of 1, the singleton of each integer and that of each rational number – all these are interpretable in (S, P). The usual ordering $\leqslant$ of the reals is intepretable in (S, P) since, for any two real numbers x, y, we have $x \leqslant y$ if and only if there is a real number t such that $x+t^2=y$. Similarly, the unary relation equal to $+$ for the positive (or nonnegative) reals is interpretable in (S, P).

The singleton of every algebraic number is interpretable in (S, P). Indeed, an algebraic number is defined as the i-th solution (in the usual ordering $\leqslant$) of a certain polynomial equation

$$a_0+a_1x+a_2x^2+\cdots a_nx^n \qquad (i \leqslant n \text{ integers}),$$

where every coefficient a is an integer. Since the degree n of the polynomial is fixed, we do not need the exponential relation, but only the sum and product of real numbers, in order to define x by means of a logical formula.

Now consider the unary relation E on the reals, equal to $+$ for the integers. The relation equal to $+$ for the natural numbers is interpretable in (S, P, E). The same holds for the relation "x, y are natural numbers and $x \leqslant y$", the relation "x, y are natural numbers and x divides y", the relation

"x, y, z are natural numbers and z is the remainder upon division of x by y", and finally the relation "x, y, z are natural numbers and $z=x$ to the power y". In general, all the arithmetical relations in the sense of 7.2, extended trivially to the entire set of reals, are interpretable in (S, P, E) by the method of 7.2.1, proposition (2′).

7.3.5. *The unary relation on the real numbers, equal to + for the algebraic numbers alone, is interpretable in* (S, P, E) (sum, product, set of integers).

▷ With each triple (n, u, v) of natural numbers, we associate a sequence of remainders $a_i (i=0, 1, \ldots, n)$ defined by the divisions $u/((i+1)\, v+1)$, and recall that every finite sequence of natural numbers is representable in this way (7.2.1, proposition (2′)). Now consider a finite sequence of integers a_i $(i=0, 1, \ldots, n)$; we shall represent this sequence by a quintuple of natural numbers (n, s, t, u, v), as follows. The triple (n, u, v) will represent the sequence of absolute values $|a_i|=$remainder of $u/((i+1)\, v+1)$. The triple (n, s, t) represents the sequence of signs of the a_i's, with the following convention: if s is divisible by $(i+1)\, t+1$, then a_i is positive or zero, and defined therefore by $a_i=$remainder of $u/(i+1)\, v+1$; otherwise, a_i is negative or zero, therefore defined by a_i+remainder of $u/((i+1)\, v+1)=0$. We see that every quintuple defines a finite sequence of integers and, conversely, every finite sequence of integers is representable by a quintuple. Moreover, the following 7-ary relation is interpretable in (S, P, E): "a is an integer, i, n, s, t, u, v are natural numbers, $i \leqslant n$, and a is the i-th number defined by remainders as above from the quintuple n, s, t, u, v".

It is convenient to define an algebraic number x from the start as a nonnegative algebraic number or minus a nonnegative algebraic number. A nonnegative real number is algebraic if and only if there exist a natural number n and a sequence of integers a_i $(i=0, 1, \ldots, n)$ such that x is a solution of the polynomial equation $a_0+a_1x+a_2x^2+\cdots+a_nx^n=0$. An alternative condition is that if a rational number p/q, whose numerator p and denominator q are positive integers, tends to x, then the expression

$$a_0+a_1p/q+a_2p^2/q^2+\cdots+a_np^n/q^n$$

tends to zero. In other words, for any positive real α there exists a positive real β such that, for any positive integers p, q such that $p \leqslant qx \leqslant p+q\beta$, we have the inequalities

$$0 \leqslant a_0q^n+a_1q^{n-1}p+a_2q^{n-2}p^2+\cdots+a_np^n+q^n\alpha \leqslant 2q^n\alpha.$$

Finally, a nonnegative real number x is algebraic if and only if there exists a quintuple of natural numbers (n, s, t, u, v) such that, for any positive α, there exists a positive number β such that, for all positive integers p, q satisfying the inequalities $p \leqslant qx \leqslant p + q\beta$, there is a quadruple of natural numbers (s', t', u', v') with the following property. The 0-th term b_0 defined by remainders from (n, s', t', u', v') is equal to the product of the 0-th term a_0 of (n, s, t, u, v) and q^n. For each $i \leqslant n-1$, the i-th term b_i of (n, s', t', u', v') plus the product of the $(i+1)$-th term a_{i+1} of (n, s, t, u, v) by $q^{n-i-1}p^{i+1}$ gives the $(i+1)$-th term b_{i+1} of (n, s', t', u', v'). Finally, the n-th term b_n of (n, s', t', u', v'), added to $q^n\alpha$, gives a nonnegative real number smaller than $2q^n\alpha$. ◁

We shall prove in Volume 2, Section 2.6.4, that the restriction (S′, P′) of (S, P) to the set of algebraic numbers is logically equivalent to (S, P) on the reals (this result was mentioned in 5.6). It will follow that the unary relation on the real numbers defined as + for the algebraic numbers alone is not interpretable in (S, P). For if A were a formula interpreting this relation, then $\underset{x}{\forall} \mathrm{A}(x)$ would take the value + for (S′, P′) and the value − for (S, P). Hence it follows from the preceding statement that the *unary relation* E (set of integers) *is not intepretable in* (S, P) (sum and product of real numbers); this result is due to [TAR, 1940, 1951].

7.4. Finitely axiomatizable multirelation

Let P be a bound logical formula and M a multirelation assignable to P. P is said to be a *saturated axiom* of M if P(M) = + and any multirelation M′ such that P(M′) = + is logically equivalent to M. M is said to be *finitely axiomatizable* if there exists a saturated axiom of M, i.e., a logical class which contains only logical equivalents of M.

Any multirelation over a finite base is finitely axiomatizable. In fact, we saw in 5.4 that a multirelation over a finite base and its isomorphic images, which are also its logical equivalents, constitute a logical class.

7.4.1. A chain or total ordering is said to be *dense* if, for any two different elements a, b of the base, there is always a third element between a and b.

Every dense chain is finitely axiomatizable.

▷ The base is either a single element or is infinite; consider the latter case. We divide dense chains into four classes, characterized as follows: the chain has neither minimal nor maximal element, it has a minimal and no maximal element, a maximal and no minimal element, or both minimal and maximal element. Each of these classes is a logical class. For example, take the conjunction of a formula defining the class of chains (see 5.4), with a binary predicate ρ, and the formula defining denseness:

$$\underset{x,\,y}{\forall}(\rho xy \wedge x \not\equiv y) \Rightarrow \underset{z}{\exists}(\rho xz \wedge \rho zy \wedge z \not\equiv x \wedge z \not\equiv y)$$

and the two formulas $\underset{x}{\forall}\underset{y}{\exists} x \not\equiv y \wedge \rho yx$ (no minimal element) and $\underset{x}{\forall}\underset{y}{\exists} x \not\equiv y \wedge \rho xy$ (no maximal element).

Each of the four classes is a logical equivalence class. In fact, we know, for example, that denumerable dense chains with neither minimal nor maximal element are all isomorphic to the chain of rational numbers (see Chapter 3, Exercise 1), so that they are all logically equivalent. Moreover, by the denumerable-model theorem (5.6.2) any dense infinite chain without minimal or maximal element has a denumerable restriction which is logically equivalent to it; this restriction is thus dense, with neither minimal nor maximal element, and is therefore isomorphic to the chain of rational numbers. $\lhd$

7.4.2. Recall that a (partial) ordering is called a *lattice* (see 2.1.3) if for all a, b there exist two elements $a \wedge b$ and $a \vee b$ such that the elements $\leqslant a$ and $\leqslant b$ are exactly those $\leqslant a \wedge b$, and the elements $\geqslant a$ and $\geqslant b$ are exactly those $\geqslant a \vee b$. The lattice is said to be *Boolean* [or a Boolean algebra] if it has a minimal element $-$ and maximal element $+$, and for every element a there is a negation $\neg a$ such that $a \wedge \neg a = -$ and $a \vee \neg a = +$; moreover, $\wedge$ and $\vee$ satisfy both distributive laws.

An element a of the base is called an *atom* of the lattice if $a \neq -$ and there is no element between $-$ and a.

There exists a denumerable Boolean lattice without atoms: consider the sets of irrational numbers in the interval (0, 1) which are finite unions of intervals with rational endpoints. Order these sets by inclusion, with the empty set playing the role of $-$, the entire interval that of $+$; the operations $\wedge$, $\vee$, $\neg$ are intersection, union, complement.

All denumerable Boolean lattices without atoms are isomorphic. This will follow from the propositions (1) and (2) below and from the fact that dense denumerable chains with neither minimal nor maximal element are isomorphic.

(1) *If a, b are two elements of the base of the lattice such that $a<b$, there exists an element c such that $a<c<b$.*

▷ By assumption, there exists an element i between $-$ and $b\wedge\neg a$ (the latter element is not $-$, since $\neg b\vee a\neq +$ or $b\nleqslant a$), such that $i\neq -$ and $i<b\wedge\neg a$. Then $a\vee i$ satisfies the inequality $a\leqslant a\vee i\leqslant b$. Moreover, $a\vee i\neq a$, since otherwise $i\leqslant a$ and $i<b\wedge\neg a$, so that $i=-$, a contradiction. Moreover, $a\vee i\neq b$, since otherwise $\neg b\vee a\vee i=+$, and so

$$b\wedge\neg a\leqslant i<b\wedge\neg a,$$

a contradiction. ◁

(2) *There exists a chain* A, *which is a restriction of the lattice, such that every element of the lattice is a Boolean combination of finitely many elements of* A (this statement is valid for any finite or denumerable Boolean lattice).

▷ Since the elements of the lattice can be indexed, it suffices to prove the following statement: If B is a finite chain, a restriction of the lattice, containing the minimal element $-$ and maximal element $+$ of the lattice, and if u is any element of the lattice, then there exists a chain B_u which is also a restriction of the lattice and an extension of B, such that u is a Boolean combination of elements of B_u.

Let $a_0=-, a_1, a_2, \ldots, a_h=+$ be the sequence of elements of B in increasing order. The element $a_1\wedge u$ lies between $-$ and a_1. Suppose that $a_n\wedge u$ is a Boolean combination of $a_0, a_1, \ldots, a_n$ and of intermediate elements; then the same holds for $a_{n+1}\wedge u$. Indeed, by hypothesis, $a_n\wedge\neg(a_{n+1}\wedge u)$ is a Boolean combination of $a_0, a_1, \ldots, a_n$ and of intermediate elements. Add the element $a_n\vee(a_{n+1}\wedge u)$, which lies between a_n and a_{n+1}. The conjunction of $\neg a_n\vee(a_{n+1}\wedge u)$ and the preceding element gives $a_{n+1}\wedge u$, and this has been done by interpolating elements only between $-$ and a_{n+1}. Since $a_n\wedge u$ is obtained by interpolating elements between $-$ and a_n, we see that $\neg a_n\wedge u$ is obtained by interpolating elements between $-$ and $\neg a_n$, i.e., between a_n and $+$. Thus $u=(a_n\wedge u)\vee(\neg a_n\wedge u)$ is obtained by interpolating elements between $-$ and $+$. ◁

Note that there exist chains which, though they are maximal restrictions of denumerable Boolean lattices without atoms, do not generate the entire lattice by Boolean combinations: Consider again the intervals of irrational numbers with rational endpoints; let e be an irrational number ($0<e<1$) and consider a maximal family of nested intervals with rational endpoints, all containing e (this example is due to R. Bonnet).

Every Boolean lattice without atoms is finitely axiomatizable, for each of the conditions which define it is expressed by a logical formula. Moreover, any two such lattices are logically equivalent to a denumerable lattice of the same type, by Theorem 5.6.2, and denumerable lattices are isomorphic, as proved above (see [LOS, 1954]).

Note that the set of finite and denumerable Boolean lattices is well ordered by embeddability (see 3.2.3) [MAY-PIE, 1960].

7.4.3. *Examples of multirelations which are not finitely axiomatizable.*

(1) No infinite set is finitely axiomatizable. In fact, the class of infinite sets is a logical equivalence class (see 5.6), but it is not a logical class; see the enumeration of logical classes of sets in 5.5.2.

(2) Consider the successor relation C on the natural numbers: $C(x, y) = +$ when $y = x+1$; we shall see in Volume 2, Section 1.6.3, that C is not finitely axiomatizable: for any logical formula P satisfied by C, one can extend C to a relation C* which has a cyclic restriction to p elements $a_1, \ldots, a_p$ ($C(a_1, a_2) = C(a_2, a_3) = \cdots = C(a_p, a_1) = +$): for sufficiently large p, P still satisfies C*, which is however not logically equivalent to C.

7.4.4. *Let* N *be a multirelation interpretable in* M; *then if* M *is finitely axiomatizable,* MN *is finitely axiomatizable, and conversely.*

▷ If M is finitely axiomatizable, consider the conjunction of a saturated axiom of M with the biconditional $\forall_{1,\ldots,n} (\sigma x^1 \ldots x^n \Leftrightarrow \mathrm{P})$, where P is the formula interpreting N in M. Conversely, if MN is finitely axiomatizable, we use P to replace σ in a saturated axiom of MN ◁

Note that if N is interpretable in M and M is finitely axiomatizable, then N is not necessarily finitely axiomatizable. For example, the successor relation C on the natural numbers is interpretable in the chain I of natural numbers, but we shall see in Volume 2, Sections 1.6.1–1.6.4, that I, therefore also (I, C), is finitely axiomatizable whereas C is not.

7.4.5. *If each of two multirelations* M *and* N *is interpretable by the other and* M *is finitely axiomatizable, then* N *is finitely axiomatizable.*

▷ Let $\mathscr{P}_1, \ldots, \mathscr{P}_h$ be the operators which interpret M by N and $\mathscr{Q}_1, \ldots, \mathscr{Q}_l$ those which interpret N by M; then

$$M = \mathscr{P}_1(N) \ldots \mathscr{P}_h(N) \quad \text{and} \quad N = \mathscr{Q}_1(M) \ldots \mathscr{Q}_l(M).$$

Let $\mathscr{A}$ be the logical class of logical equivalents of M. For every multirelation N′ of the same arity as N, consider the two conditions

$$\mathscr{P}_1(N') \ldots \mathscr{P}_h(N') \in \mathscr{A}$$

and

$$N' = \mathscr{Q}_1(\mathscr{P}_1(N') \ldots \mathscr{P}_h(N')) \ldots \mathscr{Q}_l(\mathscr{P}_1(N') \ldots \mathscr{P}_h(N')).$$

Note that each of these conditions defines a logical class: the first, because the composite operator of the operators $\mathscr{P}$ by $\mathscr{A}$ is logical, by 5.2.2; the second, because, if $\mathscr{R}_1, \ldots, \mathscr{R}_l$ denote the composite logical operators which figure therein and $\mathscr{I}_1, \ldots, \mathscr{I}_l$ the selectors which transform each N′ into itself, which are also logical, then the equality has the following meaning: the relations $\mathscr{R}_1, \ldots, \mathscr{R}_l, \mathscr{I}_1, \ldots, \mathscr{I}_l$ of N′ are pairwise identical; the first is identical to the $(l+1)$-th, …, the l-th to the $2l$-th. Now the class of multirelations in which two relations of given ranks are identical is a logical class (5.4.1), and the same holds for its inverse image under logical operators (5.4.2). The conjunction of these two conditions for N′ is also a logical class; it is sufficient to show that it implies, and is therefore equivalent to, logical equivalence to N. In fact, the first condition implies that $\mathscr{P}_1(N') \ldots \mathscr{P}_h(N')$ is logically equivalent to M, and the second that N′ is logically equivalent to N. ◁

7.5. Interpretability Theorem [BET, 1953; CRA, 1957]

Let P be a bounded formula whose predicates can be divided into two sequences, corresponding to a μ-ary multirelation and a ν-ary multirelation.

Theorem. *Assume that for any μ-ary multirelation* M *and ν-ary multirelations* N *and* N′ *with the same base as* M, *the conditions* $P(M, N) = P(M, N') = +$ *imply* $N = N'$. *Then there exists a finite sequence of logical formulas* $Q_1, \ldots, Q_h$, *whose arities constitute the sequence* ν, *such that*

$$N = Q_1(M) \ldots Q_h(M).$$

for any μ-ary multirelation M *and ν-ary multirelation* N *such that* P(M, N) = +.

Note that the same formulas Q are valid for all multirelations M and N such that P(M, N) = +; and, by the preceding assertion, the latter condition implies that N is interpretable in M.

▷ For every, say, n-ary predicate σ of arity ν, let σ' denote another n-ary predicate and P′ the formula obtained from P when each σ is replaced by σ'; we have the deduction

$$P \wedge P' \vdash \underset{1 \ldots n}{\forall} \sigma x^1 \ldots x^n \Leftrightarrow \sigma' x^1 \ldots x^n.$$

We may omit the quantifier $\forall$ in the second term, so that

$$(P \wedge \sigma x^1 \ldots x^n) \vdash (P' \Rightarrow \sigma' x^1 \ldots x^n).$$

We may assume that each term is replaced by an equideducible prenex formula, σ is inactive in the free part of the second term and σ' inactive in the free part of the first term. By the interpolation theorem 6.7.2, there exists a formula Q, whose free part contains neither σ nor σ' as active predicates, such that

$$(P \wedge \sigma x^1 \ldots x^n) \vdash Q \vdash (P' \Rightarrow \sigma' x^1 \ldots x^n).$$

The first deduction implies

$$P \vdash (\sigma x^1 \ldots x^n \Rightarrow Q).$$

From the second deduction, replacing σ' by σ in both $\sigma' x^1 \ldots x^n$ and P′, we get

$$Q \vdash (P \Rightarrow \sigma x^1 \ldots x^n),$$

or

$$(P \wedge Q) \vdash \sigma x^1 \ldots x^n,$$
$$P \vdash (Q \Rightarrow \sigma x^1 \ldots x^n).$$

Finally,

$$P \vdash (\sigma x^1 \ldots x^n \Leftrightarrow Q),$$

and, since P is bounded,

$$\mathrm{P}\vdash\underset{1\ldots n}{\forall}(\sigma x^1\ldots x^n\Leftrightarrow \mathrm{Q}),$$

which is one form of the desired conclusion. ◁

7.5.1. Given a logical formula P and a multirelation M such that, for all N and N′ of convenient arity, the equalities P(MN) = P(MN′) = + imply N = N′, it does not necessarily follow that N is interpretable by M.

For example, consider the chain I of natural numbers and the successor relation C on this set, and let P be the conjunction of the formulas

$$\underset{x}{\forall}\iota xx;\qquad \underset{xy}{\forall}\gamma xy\Rightarrow\iota xy;$$
$$\underset{xyz}{\forall}(\iota xy\wedge\iota yz)\Rightarrow\iota xz;\qquad \underset{xy}{\forall}(\iota xy\wedge\iota yx)\Rightarrow x\equiv y.$$

For any binary relation I′ on the natural numbers, if P(CI′) = +, then I′ = I, since I′ is reflexive, transitive, and antisymmetric, and any two equal or consecutive integers give it the value +. Nevertheless, we know that I is not interpretable by C (see Volume 2, Section 1.7.2).

EXERCISES

1

Consider the following relations on the natural numbers (including zero):
Zero: $Z(x) = +$ for $x = 0$;
Smaller than: $I(x, y) = +$ for $x \leqslant y$;
Between: $E(x, y, z) = +$ when $x \leqslant z \leqslant y$ or $y \leqslant z \leqslant x$.

Integral part of quotient: $Q(x, y, z) = +$ when $y \neq 0$ and z is the largest integer such that $yz \leqslant x$.

(1) Using predicates replaceable by Z, I, E, Q, find logical formulas which interpret Z by I, Z by E, Z by Q, E by I, I by E, I by Q, E by Q.

(2) Considering a permutation of the set of natural numbers, show that I, E, Q are not interpretable by Z. Let Z′, I′, E′ denote the relations defined in the same way as Z, I, E but over the integers (positive, zero, and negative). Show by considering permutations of the set of integers that Z′ is not interpretable by I′; neither are the following interpretability relations possible: I′ by Z′, E′ by Z′, Z′ by E′, I′ by E′ (though E′ is evidently interpretable by I′).

2

Using the axiom of choice, or assuming only that every infinite set is equipollent to the set of m-tuples of its elements for every natural number m, show that for every relation R there is a multirelation M on the same base, consisting of at most binary relations, such that R is interpretable by M. Compare this with the negative result of 4.2.1 for free interpretability.

It has been proved that any denumerable multirelation is interpretable by a symmetric binary relation (see [CHU-QUI, 1952]).

3

Given two bound formulas P and Q, we say that Q is *subdeducible* from P if, for any multirelation X such that $P(X) = +$, there exists a multirelation Y interpretable in X such that $Q(Y) = +$. If Q is deducible from P, then Q is subdeducible from P (take $Y = X$). Note that subdeduction is a pre-ordering (reflexive and transitive relation). If an antithesis is subdeducible from P, then P itself is an antithesis.

(1) Show that P is subdeducible from a thesis (and hence from any formula) if and only if $P(X) = +$ for at least one constant multirelation X on each base (i.e., a multirelation which is invariant under any permutation of its base; see Chapter 4, Exercise 1). Examples: $\underset{x}{\forall} \rho x$, or also $\underset{x}{\forall} \neg \rho x$. Let Q and R denote these two formulas. Show that Q and R may be both subdeducible from P while the conjunction $Q \wedge R$ is not subdeducible from P. On the other hand, if R is subdeducible from both P and Q, then R is subdeducible from the disjunction $P \vee Q$. There exists a formula P such that neither P nor $\neg P$ is subdeducible from a thesis; for instance, if p is an integer, the formula P which is true for the sets of cardinality p. Again taking $P = \underset{x}{\forall} \rho x$, show P and its negation may be subdeducible from each other.

(2) Show that if Q is subdeducible from P, then $\neg P \vee Q$ is subdeducible from a thesis. However, $P = \underset{x}{\exists} \rho x$ is consistent, and thus no antithesis Q is subdeducible from P, whereas $\neg P \vee Q$ is equideducible from $\underset{x}{\forall} \neg \rho x$ and therefore subdeducible from a thesis. It may happen that Q is subdeducible from P without $\neg P$ being subdeducible from $\neg Q$: let P be a thesis and $Q = \underset{x}{\forall} \rho x$. It may happen that $\neg P \vee Q$ takes the value $+$ for every constant multirelation assignable to it, while Q is not subdeducible from P: let P be the formula $\underset{x}{\exists} \rho x \wedge \underset{x}{\exists} \neg \rho x$ and Q an antithesis.

(3) We say that Q is *quasideducible* from P if Q is subdeducible from P and $\neg$P is subdeducible from $\neg$Q. If Q is deducible from P, then Q is quasideducible from P. Quasideduction is a pre-ordering. A formula quasideducible from a thesis must be a thesis; an antithesis can be quasideducible only from an antithesis. The formula $\underset{x}{\forall}\rho x$ and its negation are quasideducible from one another.

(4) A formula A is a thesis if and only if, for all P and Q, if $\neg P \vee Q$ is deducible from A, then Q is subdeducible from P (J.-C. Collet).

REFERENCES

The following list includes (1) books or articles referred to in the text, indicated there by the first letters of the author's name and the year of publication, and (2) certain classical works of logic which may be consulted along with the present course (in particular, the literature mentioned in the Introduction).

The solutions to the Exercises of Chapters 1 and 2, by J.-P. Bénéjam and E. François, have been published by l'Institut Blaise Pascal, Paris, 1963. The solutions of all exercises of volume 1, by M. H. Dulac and J. Milon are in press by Gauthier-Villars, Paris.

Aanderaa, S., P. Andrews and B. Dreben

1963: 'False Lemmas in Herbrand', *Bull. Am. Math. Soc.* **69**, No. 5, 699–706.

Ax, J. and S. Kochen

1965: 'Diophantine Problems over Local Fields', *Am. J. Math.* **87**, 605–648.

1966: 'Diophantine Problems over Local Fields: III. Decidable Fields', *Ann. Math.* **83**, 437–456.

Bénéjam, J.-P.

1969: 'Applications du théorème de Herbrand à la présentation de thèses tératologiques du calcul des prédicats élémentaire', *Comptes Rendus* **268(A)**, 757–60.

1970: 'Une remarque sur des théorèmes de Rado et Fraïssé', *Comptes Rendus* **270(A)**, 1656–58.

Bernays, P.

1926: 'Axiomatische Untersuchung des Aussagenkalküls der Principia Mathematica', *Math. Z.* **25**, 305–20.

1937–1954: 'A System of Axiomatic Set Theory', *J. Symbolic Logic* **2** (1937) 65–77; **6** (1941) 1–17; **7** (1942) 65–89, 133–45; **8** (1943) 89–106; **13** (1948) 65–79; **19** (1954) 81–96.

1958: *Axiomatic Set Theory*, with Introduction by A. A. Fraenkel, North-Holland Publ. Co., Amsterdam.

Beth, E. W.

1951: *Les fondements logiques des mathématiques*; 2nd ed., Gauthier-Villars, Paris, 1955.

1953: 'On Padoa's Method in the Theory of Definition', *Proc. Kon. Nederl. Akad. Wetens.* **A 56**, No. 4, and *Indag. Math.* **15**, No. 4, 330–39.

1959: *The Foundations of Mathematics*, North-Holland Publ. Co., Amsterdam.

Boole, G.

1847: *The Mathematical Analysis of Logic, Being an Essay Toward a Calculus of Deductive Reasoning*, Cambridge and London; reprinted Oxford and New York, 1948.

1854: *An Investigation of the Laws of Thought, on which Are Founded the Mathematical Theories of Logic and Probabilities*, London; reprinted as Vol. 2 of Boole's *Collected Works*, Chicago and London, 1916; reprinted New York, 1951.

Bourbaki, N.

1954: *Éléments de mathématique*, Livre I, *Théorie des ensembles*, Chap. 1 and 2, Hermann, Paris.

Calais, J.-P.
1967: 'Relation et multirelation pseudohomogènes', *Comptes Rendus* **265(A)**, 2–4.
Carnap, R.
1934: *Die logische Syntax der Sprache*, Springer, Vienna; English translation by A. Smeaton: *The Logical Syntax of Language*, New York and London, 1937.
Church, A.
1936: 'A Bibliography of Symbolic Logic', *J. Symbolic Logic* **1**, No. 4.
1936a: 'An Unsolvable Problem of Elementary Number Theory', *Am. J. Math.* **58**, 345–63.
1956: *Introduction to Mathematical Logic*, Vol. I, Princeton University Press.
Church, A. and W. V. O. Quine
1952: 'Some Theorems on Definability and Decidability', *J. Symbolic Logic* **17**, 179–87.
Cohen, J.
1957: 'Can the Logic of Indirect Discourse Be Formalized?', *J. Symbolic Logic* **22**, 225–32.
Cohen, P.
1963: 'The Independence of the Continuum Hypothesis', *Proc. Nat. Acad. Sci. U.S.A.* **50**, 1143–48; **51**, 105–10.
1966: *Set Theory and the Continuum Hypothesis*, Benjamin, New York.
Craig, W.
1957: 'Linear Reasoning. A New Form of the Herbrand-Gentzen Theorem', *J. Symbolic Logic* **22**, 250–68.
Curry, H. B.
1952: *Leçons de logique algébrique*, Gauthier-Villars, Paris.
Cusin, R. and J.-F. Pabion
1970: 'Une généralisation de l'âge des relations', *Comptes Rendus* **270 (A)** 17–20.
Dedekind, R.
1888: *Was sind und was sollen die Zahlen?*, Braunschweig.
Dopp, J.
1950: *Leçons de logique formelle*, Louvain.
Dushnik, B. and E. W. Miller
1940: 'Concerning Similarity Transformations of Linearly Ordered Sets', *Bull. Am. Math. Soc.* **46**, 322–26.
Erdös, P. and R. Rado
1950: 'A Combinatorial Theorem', *J. London Math. Soc.* **25**, 249–55.
Fraenkel, A, A,
1925: 'Untersuchungen über die Grundlagen der Mengenlehre', *Math. Z.* **22**, 250–73.
Fraïssé, R.
1954: 'Sur l'extension aux relations de quelques propriétés des ordres', *Ann. E.N.S.* **71**, 363–88.
1969: 'Réflexions sur les axiomatiques de l'arithmétique, l'algèbre et la géométrie élémentaires, *L'âge de la science*, 192–239.
1971: 'Abritement entre relations, et spécialement entre chaînes', *Symposia Math. Ist. Naz. Alta Mat. Roma*, Vol. 5, pp. 203–51.
Frasnay, C.
1965: 'Quelques problèmes combinatoires concernant les ordres totaux et les relations monomorphes', *Ann. Inst. Fourier (Grenoble)* **15**, No. 2, 415–524.
Frege, G.
1879: *Eine Begriffsschrift, der arithmetischen machgebildeten Formelsprache des reinen Denkens*, Nebert, Halle.

1891: 'Funktion und Begriff', Address given to the Jenaische Gesellschaft für Medizin und Naturwissenschaft.
1892: 'Über Sinn und Bedeutung', *Z. Phil. Philos. Kritik* **100**, 25–50.
English translations: M. Black, *Phil. Rev.* **57** (1948) 207–30;
H. Feigl, *Readings in Philosophical Analysis*, pp. 85–102, 1949;
Review in *J. Symbolic Logic* **13** (1948) 152 and **14** (1949) 184.

Gentzen, G.
1934: 'Untersuchungen über das logische Schliessen', *Math. Z.* **39**, 176–210 and 405–31.

Gleyzal, A.
1940: 'Order Types and Structure of Orders. I', *Trans. Am. Math. Soc.* **48**, 451–466.

Gödel, K.
1930: 'Die Vollständigkeit der Axiome des logischen Funktionenkalküls', *Monatsh. Math. Phys.* **37**, 349–60.
1931: 'Über formal unentscheidbare Sätze der Principia Mathematica und verwandter Systeme', *Monatsh. Math. Phys.* **38**, 173–98.
1940: *The Consistency of the Axiom of Choice and of the Generalized Continuum Hypothesis with the Axioms of Set Theory*, Princeton University Press.

Götlind, E.
1947: 'An Axiom System for the Calculus of Propositions', *Norsk Mat. Tidsskr.* **29**, 1–4.

Hatcher, W.
1968: *Foundations of Mathematics*, Saunders, Philadelphia.

Henkin, L.
1949: 'Fragments of the Propositional Calculus', *J. Symbolic Logic* **14**, 42–48.
1949a: 'The Completeness of the First-Order Functional Calculus', *J. Symbolic Logic* **14**, 159–66.
1953: 'Some Interconnections Between Modern Algebra and Mathematical Logic', *Trans. Am. Math. Soc.* **74**, 410–27.

Herbrand, J.
1930: *Recherches sur la théorie de la démonstration*, thèse, Paris.
1968: *Écrits logiques* (ed. by J. van Heijenoort), P.U.F., Paris.

Heyting, A.
1930: 'Die formalen Regeln der intuitionistischen Logik', *S.-B. Preuss. Akad. Wiss., Phys.-Math. Kl.*, 42–56.

Higman, G.
1952: 'Ordering by Divisibility in Abstract Algebras', *Proc. London Math. Soc.* (3) **2**, 326–36.

Hilbert, D.
1904: 'Über die Grundlagen der Logik und der Arithmetik', *Proc. Third Internat. Math. Congr.* 174–85.

Hilbert, D. and W. Ackermann
1928: *Grundzüge der theoretischen Logik*; 2nd ed., 1938; 3rd ed., 1949, Springer Verlag, Berlin.
1950: *Principles of Mathematical Logic* (English translation of 2nd edition of above), Chelsea, New York.

Hilbert, D. and P. Bernays
1934: *Grundlagen der Mathematik*, Vol. 1, Springer, Berlin; reprinted, Edwards, Ann Arbor, Mich., 1944.
1939: *Ibid.*, Vol. 2., Springer, Berlin; reprinted, Edwards, Ann Arbor, Mich., 1944.

Jaśkowski, S.
1948: 'Un calcul des propositions pour les systèmes déductifs', *Studia Soc. Sci. Torunenseis* **1**, No. 1, 57–77.

Jean, M.
1967: 'Relations monomorphes et classes universelles', *Comptes Rendus* **264**, 591–93.

Jonsson, B.
1965: 'Extensions of Relational Structures', in: *Theory of Models*, Proc. Intern. Symp. Berkeley, North-Holland Publ. Co., Amsterdam, pp. 146–57.

Kalmár, L.
1935: 'Über die Axiomatisierbarkeit des Aussagenkalküls', *Acta Sci. Math. (Szeged)* **7**, 222–43.

Keisler, J.
1960: 'Theory of Models with Generalized Atomic Formulas', *J. Symbolic Logic* **25**, 1–26.

Kleene, S. C.
1952: *Introduction to Metamathematics*, North-Holland Publ. Co., Amsterdam.

Kuratowski, C.
1920: 'Sur la notion de l'ensemble fini', *Fund. Math.* **1**, 129–31.

Laver, R.
1971: 'On Fraïssé's Order Type Conjecture', Thesis, Berkeley, 1969, and *Ann. Math.* **93**, 89–111.

Lewis, C. I.
1918: *A Survey of Symbolic Logic*, University of California Press, Berkeley.

Łos, J.
1954: 'On the Categoricity in Power of Elementary Deductive Systems and Some Related Problems', *Colloq. Math.* **3**, No. 1, 58–62.
1955: 'Quelques remarques, théorèmes et problèmes sur les classes définissable d'algèbres', in: *Mathematical Interpretation of Formal Systems*, North-Holland Publ. Co., Amsterdam, pp. 98–113.

Löwenheim, L.
1915: 'Über Möglichkeiten im Relativkalkül', *Math. Ann.* **76**, 447–70.

Łukasiewicz, J.
1921: 'Logika dwuwartościowa' ('Two-Valued Logic'), *Przeglad Filozoficzny* **23**, 189–205.
1929: *Elementy logiki matematyczny*, Warsaw (mimeographed).
1948: 'The Shortest Axiom of the Calculus of Propositions', *Proc. Roy. Irish Acad.*, Sect. A **52**, 25–33.

Łukasiewicz, J. and A. Tarski
1930: 'Untersuchungen über den Aussagenkalkül', *Comptes Rendus Soc. Sci. Lett. Varsovie, Classe* III **23**, 30–50.

Lyndon, R. C.
1951: 'Identities in Two-Valued Calculi', *Trans. Am. Math. Soc.* **71**, 457–465.
1959: 'An Interpolation Theorem in the Predicate Calculus', *Pacific J. Math.* **9**, 129–42.

Mal'cev, A. I.
1936: 'Untersuchungen aus dem Gebiete der mathematischen Logik', *Rec. Math. (Mat. Sbornik), Nouv. Sér.* **1**, 323–36.

Malitz, J.
1967: 'On Extending Multirelations', *Notices Am. Math. Soc.* **14**, No. 6, 828 (Abstract 67 T 577).

Mayer, R. D. and R. S. Pierce
1960: 'Boolean Algebras with Ordered Bases', *Pacific J. Math.* **10**, 925–42.
Mendelson, E.
1964: *Introduction to Mathematical Logic*, Van Nostrand, Princeton, N.J.
de Morgan, A.
1847: *Formal Logic*; reprinted, Taylor, London, 1926.
Mostowski, A.
1939: 'Über die Unabhängigkeit des Wohlordnungssatzes vom Ordnungsprinzip', *Fund. Math.* **32**, 201–51.
von Neumann, J.
1929: 'Über ein Widerspruchsfreiheitsfrage in der axiomatischen Mengenlehre', *J. Reine Angew. Math.* **160**, 227–41.
Nicod, J. G. P.
1917: 'A Reduction in the Number of the Primitive Propositions of Logic', *Proc. Cambridge Philos. Soc.* **19**, 32–41.
Padoa, A.
1902: 'Un nouveau système irréductible de postulats pour l'algèbre', *Comptes Rendus 2ème Congr. Intern. Math.*, pp. 249–56.
Pastel, Anne-Marie
1971: 'Sur les fonctions logiques permutantes', *Comptes Rendus* **272** (**A**), 1154–56.
Peano, G.
1894–1908: *Formulaire de mathématiques* (introduction and five volumes, edited by Peano and written by him in collaboration with seven other authors), Turin.
Peirce, C. S.
1885: 'On the Algebra of Logic: A Contribution to the Philosophy of Notation', *Am. J. Math.* **7**, 180–202.
1931: *Collected Papers*, Harvard University Press, Cambridge, Mass.
Porte, J.
1958: 'Deux systèmes simples pour le calcul des propositions', *Publ. Sci. Univ. Alger. Sér. A* **5**, 5–16.
Post, E.
1921: 'Introduction to a General Theory of Elementary Propositions', *Am. J. Math.* **43**, 163–85.
1941: *The Two-Valued Iterative Systems of Mathematical Logic*, Ann. Math. Studies No. 5, Princeton University Press (analyzed by S. C. Kleene in *Math. Reviews* **2** (1941) 337, and by H. E. Vaughan in *J. Symbolic Logic* **6** (1941) 114).
Presburger, M.
1929: 'Über die Vollständigkeit eines gewissen Systems der Arithmetik ganzer Zahlen, in welchem die Addition als einzige Operation hervortritt', *Comptes Rendus 1er Congr. Math. Pays Slaves*, Warsaw, pp. 92–101.
Quine, W. V. O.
1940: *Mathematical Logic*, Norton, New York; revised edition, Harvard University Press, 1951.
1963, 1969: *Set Theory and Its Logic*, Harvard University Press, Cambridge, Mass.
Rado, R.
1949: 'Axiomatic Treatment of Rank in Infinite Sets', *Canad. J. Math.* **1**, 337–43.
Ramsey, F. P.
1926: 'The Foundations of Mathematics', *Proc. London Math. Soc.* (2) **25**, 338–84.
1929: 'On a Problem in Formal Logic', *Proc. London Math. Soc.* (2) **30**, 264–86.

Rasiowa, Helena
1949: 'Sur un certain système d'axiomes du calcul des propositions', *Norsk Mat. Tidsskr.* **31**, 1–3.
Robinson, A.
1963: *Introduction to Model Theory and to the Metamathematics of Algebra*, North-Holland Publ. Co., Amsterdam.
Robinson, Julia
1949: 'Definability and Decision Problems in Arithmetic', *J. Symbolic Logic* **14**, 98–114.
Rosenbloom, P. C.
1950: *The Elements of Mathematical Logic*, Dover, New York.
Rosser, J. B.
1953: *Logic for Mathematicians*, New York.
Russell, B.
1903: *The Principles of Mathematics*, London; 2nd ed., London, 1937 and New York, 1938.
1908: 'Mathematical Logic as Based on the Theory of Types', *Am. J. Math.* **30**, 222–62.
Schröder, E.
1895: *Vorlesungen über die Algebra der Logik*, Vol. 3, *Algebra und Logik der Relative*, Leipzig.
Schur, I.
1916: 'Über die Kongruenz $x^m + y^m = z^m \pmod p$', *Jahr. Deutsch. Math. Verein.* **25**, 114–17.
Sheffer, M. M.
1913: 'A Set of Five Independent Postulates for Boolean Algebras, with Application to Logical Constants', *Trans. Am. Math. Soc.* **14**, 481–88.
Sierpinski, W.
1918: 'L'axiome de M. Zermelo et son rôle dans la théorie des ensembles et l'analyse', *Bull. Acad. Sci. Cracovie* 97–152.
1933: 'Sur un problème de la théorie des relations', *Ann. Scuola Norm. Sup. Pisa* **2**, Ser. 2, 285–87.
Skolem, T.
1920: 'Logisch-kombinatorische Untersuchungen über die Erfüllbarkeit oder Beweisbarkeit mathematischer Sätze nebst einem Theoreme über dichte Mengen', *Skr. utgit Viden. Kristiana, I, Math.-Naturw. Klasse*, No. 4.
Sobocinski, B.
1954: 'Axiomatization of a Conjunctive-Negative Calculus of Propositions', *J. Comp. Syst.* **1**, 229–42.
Stackel, P.
1907: 'Zu H. Webers elementarer Mengenlehre', *Jahr. Deutsch. Math. Verein.* **16**, 425.
Suppes, P. S.
1957: *Introduction to Logic*, Van Nostrand, Princeton, N.J.
1960: *Axiomatic Set Theory*, Van Nostrand, Princeton, N.J.
Suszko, R.
1957: 'Formalna teoria wartości logicznych, I', *Studia Logica* **6**, 145–237.
Tarski, A.
1924: 'Sur les ensembles finis', *Fund. Math.* **6**, 45–95.
1930: 'Über einige fundamentale Begriffe der Metamathematik', *Comptes Rendus Soc. Sci. Lett. Varsovie, Classe III* **23**, 22–29 (English translation in 1956, below).
1934: *Fund. Math.* **23**, 161 (remark at end of paper by Skolem).

1936: 'Der Wahrheitsbegriff in den formalisierten Sprachen', *Studia Phil.* **1**, 261–404 (English translation in 1956, below).
1940: *The Completeness of Elementary Algebra and Geometry*, Herman, Paris.
1941: *Introduction to Logic and to the Methodology of Deductive Sciences*; 2nd ed., 1946; reprinted, 1951, New York.
1951: *A Decision Method for Elementary Algebra and Geometry*, 2nd ed., University of California, Berkeley and Los Angeles.
1952: 'Some Notions and Methods on the Borderline of Algebra and Metamathematics', *Proc. Intern. Congr. Math. Cambridge, Mass.*, 1950, Vol. 1, pp. 705–20.
1953: 'Universal Arithmetical Classes of Mathematical Systems', *Bull. Am. Math. Soc.* **59**, 390–91.
1956: *Logic, Semantics, Metamathematics*, collection of papers from 1923 to 1938 (translated by H. J. Woodger), Clarendon Press, Oxford.
1956a: *Ordinal Algebras* (with appendices by C. C. Chang and B. Jonsson), North-Holland Publ. Co., Amsterdam.

Tarski, A., A. Mostowski and R. M. Robinson
1953: *Undecidable Theories*, North-Holland Publ. Co., Amsterdam.

Vaught, R. L.
1953: 'Remarks on Universal Arithmetical Classes', *Bull. Am. Math. Soc.* **59**, 391.
1961: 'Denumerable Models of Complete Theories', in: *Infinitistic Methods* (Symposium on Foundations of Mathematics, Warsaw, 1959), pp. 303–21.
1963: 'Models of Complete Theories', *Bull. Am. Math. Soc.* **69**, 299–313.

Wajsberg, M.
1935: 'Beiträge zum Metaaussagenkalkül', *Monatsh. Math. Phys.* **42**, 221–42.

Wang, Hao
1953: 'Quelques notions d'axiomatique', *Revue Phil. Louvain* **5**, 409–43.
1963: *A Survey of Mathematical Logic*, Science Press, Peking and North-Holland Publ. Co., Amsterdam.

Wang, Hao and R. McNaughton
1953: *Les systèmes axiomatiques de la théorie des ensembles*, Gauthier-Villars, Paris.

Ward-Henson, C.
1972: 'Countable Homogeneous Relational Structures, and $\aleph_0$-Categorial Theories', *J. Symbolic Logic* **37**, 494–500.

Whitehead, A. N. and B. Russell
1910–1913: *Principia mathematica*, Vols. 1, 2, 3, Cambridge University Press.

Wittgenstein, L.
1922: *Tractatus logico-philosophicus*, Harcourt Brace, New York, and Kegan Paul, Trench, Trubner, London.

Zermelo, E.
1904: 'Beweis dass jede Menge wohlgeordnet werden kann', *Math. Ann.* **59**, 514–16.
1908: 'Untersuchungen über die Grundlagen der Mengenlehre', *Math. Ann.* **65**, 261–81.

INDEX

(The figures refer to chapter and section)

SYNTHESE LIBRARY

Monographs on Epistemology, Logic, Methodology,

Philosophy of Science, Sociology of Science and of Knowledge, and on the

Mathematical Methods of Social and Behavioral Sciences

(MARIO BUNGE (ed.), *Exact Philosophy. Problems, Tools, and Goals.* 1973, X + 214 pp.

ROBERT S. COHEN and MARX W. WARTOFSKY (eds.), *Boston Studies in the Philosophy of Science.* Volume IX: *A. A. Zinov'ev: Foundations of the Logical Theory of Scientific Knowledge (Complex Logic).* Revised and Enlarged English Edition with an Appendix by G. A. Smirnov, E. A. Sidorenka, A. M. Fedina, and L. A. Bobrova. 1973, XXII + 301 pp. Also available as a paperback.

K. J. J. HINTIKKA, J. M. E. MORAVCSIK, and P. SUPPES (eds.), *Approaches to Natural Language. Proceedings of the 1970 Stanford Workshop on Grammar and Semantics.* 1973, VIII + 526 pp. Also available as a paperback.

WILLARD C. HUMPHREYS, JR. (ed.), *Norwood Russell Hanson: Constellations and Conjectures.* 1973, X + 282 pp.

MARIO BUNGE, *Method, Model and Matter.* 1973, VII + 196 pp.

MARIO BUNGE, *Philosophy of Physics.* 1973, IX + 248 pp.

LADISLAV TONDL, *Boston Studies in the Philosophy of Science.* Volume X: *Scientific Procedures.* 1973, XIII + 268 pp. Also available as a paperback.

SÖREN STENLUND, *Combinators, λ-Terms and Proof Theory.* 1972, 184 pp.

DONALD DAVIDSON and GILBERT HARMAN (eds.), *Semantics of Natural Language.* 1972, X + 769 pp. Also available as a paperback.

MARTIN STRAUSS, *Modern Physics and Its Philosophy. Selected Papers in the Logic, History, and Philosophy of Science.* 1972, X + 297 pp.

‡STEPHEN TOULMIN and HARRY WOOLF (eds.), *Norwood Russell Hanson: What I Do Not Believe, and Other Essays,* 1971, XII + 390 pp.

‡ROBERT S. COHEN and MARX W. WARTOFSKY (eds.), *Boston Studies in the Philosophy of Science.* Volume VIII: *PSA 1970. In Memory of Rudolf Carnap* (ed. by Roger C. Buck and Robert S. Cohen). 1971, LXVI + 615 pp. Also available as a paperback.

‡YEHOSUA BAR-HILLEL (ed.), *Pragmatics of Natural Languages.* 1971, VII + 231 pp.

‡ROBERT S. COHEN and MARX W. WARTOFSKY (eds.), *Boston Studies in the Philosophy of Science.* Volume VII: *Milič Čapek: Bergson and Modern Physics.* 1971, XV + 414 pp.

‡CARL R. KORDIG, *The Justification of Scientific Change.* 1971, XIV + 119 pp.

‡JOSEPH D. SNEED, *The Logical Structure of Mathematical Physics.* 1971, XV + 311 pp.

‡JEAN-LOUIS KRIVINE, *Introduction to Axiomatic Set Theory.* 1971, VII + 98 pp.

‡RISTO HILPINEN (ed.), *Deontic Logic: Introductory and Systematic Readings.* 1971, VII + 182 pp.

‡EVERT W. BETH, *Aspects of Modern Logic.* 1970, XI + 176 pp.

‡PAUL WEINGARTNER and GERHARD ZECHA (eds.), *Induction, Physics, and Ethics, Proceedings and Discussions of the 1968 Salzburg Colloquium in the Philosophy of Science.* 1970, X + 382 pp.

‡ROLF A. EBERLE, *Nominalistic Systems.* 1970, IX + 217 pp.

‡JAAKKO HINTIKKA and PATRICK SUPPES, *Information and Inference.* 1970, X + 336 pp.

‡KAREL LAMBERT, *Philosophical Problems in Logic. Some Recent Developments.* 1970, VII + 176 pp.

‡P. V. TAVANEC (ed.), *Problems of the Logic of Scientific Knowledge.* 1969, XII + 429 pp.

‡ROBERT S. COHEN and RAYMOND J. SEEGER (eds.), *Boston Studies in the Philosophy of Science.* Volume VI: *Ernst Mach: Physicist and Philosopher.* 1970, VIII + 295 pp.

‡MARSHALL SWAIN (ed.), *Induction, Acceptance, and Rational Belief.* 1970, VII + 232 pp.

‡NICHOLAS RESCHER *et al.* (eds.), *Essays in Honor of Carl G. Hempel. A Tribute on the Occasion of his Sixty-Fifth Birthday.* 1969, VII + 272 pp.

‡PATRICK SUPPES, *Studies in the Methodology and Foundations of Science. Selected Papers from 1911 to 1969.* 1969, XII + 473 pp.

‡JAAKKO HINTIKKA, *Models for Modalities. Selected Essays.* 1969, IX + 220 pp.

‡D. DAVIDSON and J. HINTIKKA (eds.), *Words and Objections: Essays on the Work of W. V. Quine.* 1969, VIII + 366 pp.

‡J. W. DAVIS, D. J. HOCKNEY and W. K. WILSON (eds.), *Philosophical Logic.* 1969, VIII + 277 pp.

‡ROBERT S. COHEN and MARX W. WARTOFSKY (eds.), *Boston Studies in the Philosophy of Science.* Volume V: *Proceedings of the Boston Colloquium for the Philosophy of Science 1966/1968*, VIII + 482 pp.

‡ROBERT S. COHEN and MARX W. WARTOFSKY (eds.), *Boston Studies in the Philosophy of Science.* Volume IV: *Proceedings of the Boston Colloquium for the Philosophy of Science 1966/1968.* 1969, VIII + 537 pp.

‡NICHOLAS RESCHER, *Topics in Philosophical Logic.* 1968, XIV + 347 pp.

‡GÜNTHER PATZIG, *Aristotle's Theory of the Syllogism. A Logical-Philological Study of Book A of the Prior Analytics.* 1968, XVII + 215 pp.

‡C. D. BROAD, *Induction, Probability, and Causation. Selected Papers.* 1968, XI + 296 pp.

‡ROBERT S. COHEN and MARX W. WARTOFSKY (eds.), *Boston Studies in the Philosophy of Science.* Volume III: *Proceedings of the Boston Colloquium for the Philosophy of Science 1964/1966.* 1967, XLIX + 489 pp.

‡GUIDO KÜNG, *Ontology and the Logistic Analysis of Language. An Enquiry into the Contemporary Views on Universals.* 1967, XI + 210 pp.

*EVERT W. BETH and JEAN PIAGET, *Mathematical Epistemology and Psychology.* 1966, XXII + 326 pp.

*EVERT W. BETH, *Mathematical Thought. An Introduction to the Philosophy of Mathematics.* 1965, XII + 208 pp.

‡PAUL LORENZEN, *Formal Logic*. 1965, VIII + 123 pp.

‡GEORGES GURVITCH, *The Spectrum of Social Time*. 1964, XXVI + 152 pp.

‡A. A. ZINOV'EV, *Philosophical Problems of Many-Valued Logic*. 1963, XIV + 155 pp.

‡MARX W. WARTOFSKY (ed.), *Boston Studies in the Philosophy of Science*. Volume I: *Proceedings of the Boston Colloquium for the Philosophy of Science, 1961–1962*. 1963, VIII + 212 pp.

‡B. H. KAZEMIER and D. VUYSJE (eds.), *Logic and Language. Studies dedicated to Professor Rudolf Carnap on the Occasion of his Seventieth Birthday*. 1962, VI + 256 pp.

*EVERT W. BETH, *Formal Methods. An Introduction to Symbolic Logic and to the Study of Effective Operations in Arithmetic and Logic*. 1962, XIV + 170 pp.

*HANS FREUDENTHAL (ed.), *The Concept and the Role of the Model in Mathematics and Natural and Social Sciences. Proceedings of a Colloquium held at Utrecht, The Netherlands, January 1960*. 1961, VI + 194 pp.

‡P. L. GUIRAUD, *Problèmes et méthodes de la statistique linguistique*. 1960, VI + 146 pp.

*J. M. BOCHEŃSKI, *A Precis of Mathematical Logic*. 1959, X + 100 pp.

SYNTHESE HISTORICAL LIBRARY

Texts and Studies
in the History of Logic and Philosophy

LEWIS WHITE BECK (ed.), *Proceedings of the Third International Kant Congress*. 1972, XI + 718 pp.

‡KARL WOLF and PAUL WEINGARTNER (eds.), *Ernst Mally: Logische Schriften*. 1971, X + 340 pp.

‡LEROY E. LOEMKER (ed.), *Gottfried Wilhelm Leibnitz: Philosophical Papers and Letters*. A Selection Translated and Edited, with an Introduction. 1969, XII + 736 pp.

‡M. T. BEONIO-BROCCHIERI FUMAGALLI, *The Logic of Abelard*. Translated from the Italian. 1969, IX + 101 pp.

Sole Distributors in the U.S.A. and Canada:

*GORDON & BREACH, INC., 440 Park Avenue South, New York, N.Y. 10016
‡HUMANITIES PRESS, INC., 303 Park Avenue South, New York, N.Y. 10010

CPSIA information can be obtained
at www.ICGtesting.com
Printed in the USA
LVHW041114120420
653141LV00016B/1020

9 789027 702685